*In an ancient Navajo
chant of creation,
it is said that
before the sun,
the moon,
or the stars,
the mountains of this land
shone with a faint glow,
like foxfire.
From these
"shining mountains"
flashed rivers of light
spreading like lightning
across the world.
These were the "shining waters."
These are the rivers of Colorado.*

Canoeing on Ruby and Horsethief Canyon on the Colorado River. KAHNWEILER/JOHNSON.

South Fork of the Rio Grande River. JEFF GNASS.

Glenwood Canyon on the Colorado River. KAHNWEILER/JOHNSON.

San Miquel River. JEFF GNASS.

Devil's Grave Mesa on the Yampa River. JEFF GNASS.

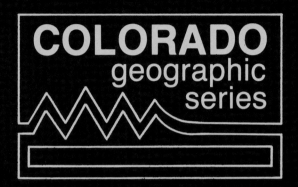

The Colorado Geographic Series

The Colorado Geographic Series is long overdue—but well worth waiting for.

The Rivers of Colorado is the first of a long series of similar portraits of Colorado's diverse and plentiful geographical features. Future books in the series will cover various natural features in the state, including the human side of Colorado—how the state's natural resources affect the lives of Coloradans.

This series is, in short, a portfolio of Colorado—and its land and people—or better yet, a celebration. With the Colorado Geographic Series, Coloradans will, finally, have something to really show off their state.

Each book will be lavishly illustrated with graphics and the highest-quality photography from America's best photographers. Each book will unfold the state's geology, history, culture, natural resources, recreational and industrial use, and scenic grandeur in a blend of graphics, photos, and text. Each book is a unique opportunity to learn more about the state of Colorado.

These books will be collector's items. Don't miss a single one.

Mike Sample and Bill Schneider, Publishers
Falcon Press Publishing Co., Inc.

The Rivers of Colorado

BY JEFF RENNICKE

NUMBER ONE
OF THE
COLORADO GEOGRAPHIC SERIES

PUBLISHED BY

Contents

Colorado, the mother of rivers	10
Restless earth, restless rivers	16
Always looking around the next bend	30
The rivers, a tale of twenty-five	48
Divided waters, divided people	90
Reflections on the never-ending story	110

The author on the Green River. JILL WOLF-RENNICKE.

Acknowledgements

Like a system of high-country creeks which eventually twine together to form a river, this book was born from many sources.

The Denver office of the U.S. Geological Survey helped decipher the complex geology of the rivers of Colorado. The Denver Public Library and Colorado Historical Society opened their vast resources to me as I traced the human history of rivers. The Denver Natural History Association and Museum provided similar aid in tracking the natural history. But much of my research was firsthand—on the river. For that, I thank Jerry Mallett and Tom Beck. Conversations during long days on the river, late-night shuttles and while watching the moon rise around campfires have, in many cases, found their way into these pages. Clifton R. Merritt of the American Wilderness Alliance shared his knowledge of the efforts to protect our last wild rivers. Editor/publisher Bill Schneider took the moonbeams, rapids and meanders and turned them into a book, and editor Russell Hill helped me improve the text in many ways. The photographers, of course, added color and life to the book, as did Biff Karlyn's graphics.

And finally, thanks to my wife Jill whose depth and understanding of rivers appears throughout the book.

The author

Jeff Rennicke is an outdoor writer whose work appears in *Sierra, New Age, Backpacker, River Runner* and other magazines. Also a correspondent to *Canoe*, contributing editor to *Wild America*, and a professional river guide, he spends his free time on rivers, and has paddled most rivers in Colorado and others above the Arctic Circle and in the Yukon. Jeff and his wife, Jill, currently live in Boulder, Colorado.

Dedication

This book is dedicated to Jerry Mallett and Tom Beck, river partners.

Colorado Geographic Series staff
Editor: Bill Schneider
Photo Editor: Mike Sample
Copy Editor: Russell Hill
Design: Bill Schneider
Graphics: Biff Karlyn
Layout: Jeanette Barnes Geary
Typesetting: Margaret Loos, Shirley Miller

Photo credits
Front Cover photos clockwise from upper left:
Columbine along a mountain stream, LARRY ULRICH;
Gates of Lodore on the Green River, GERRY WOLFE;
Rio Grande River in San Juan Range, DAVID MUENCH;
irrigated field near Westcliffe, MICHAEL WEEKS;
black-crowned night heron, GARY ZAHM;
floaters in South Canyon Rapid on the Colorado River,
DOUG LEE/TOM STACK AND ASSOCIATES.
Photo on page 1: JERRY DOWNS/THE STOCK BROCKER.

Copyright © 1985 by Falcon Press Publishing Co., Inc., Billings and Helena, Montana
All rights reserved, including the right to reproduce this book or any part thereof, in any form, except for brief quotations in a review, without written permission from the publisher.
Pre-press by Falcon Press, Helena, Montana
Manufactured by DNP America of San Francisco
Printed in Japan
ISBN: 0-934318-59-X softcover, 0-934318-68-9 cloth
Library of Congress Number: 65-80606.

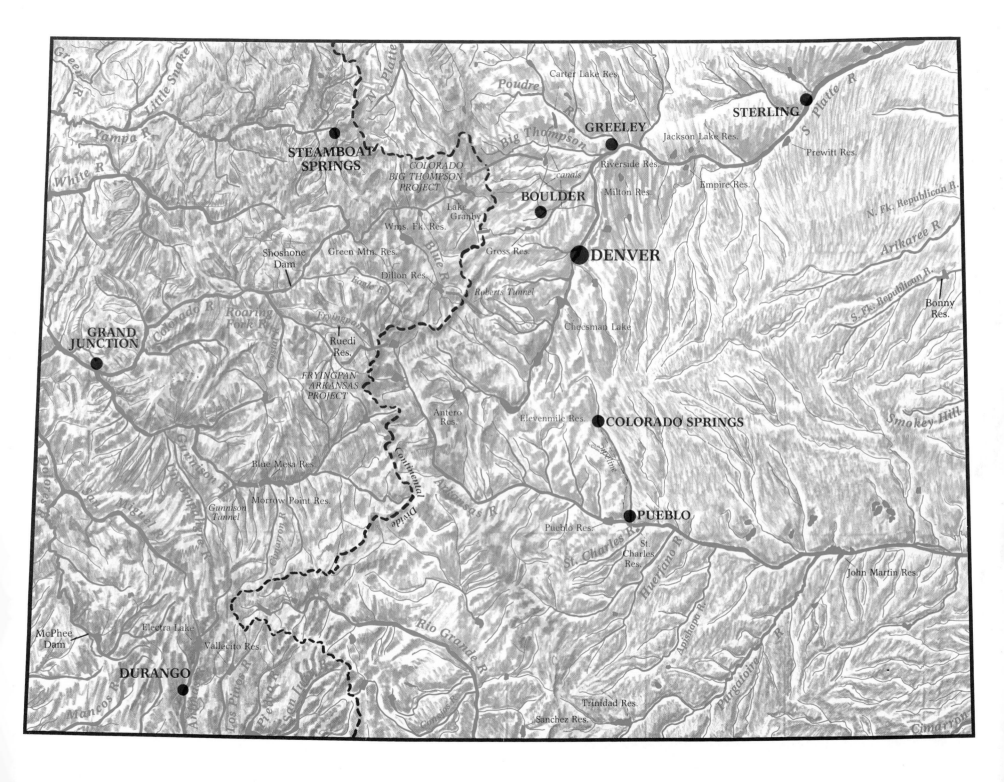

A foreword:

Colorado, the mother of rivers

Sometimes, it is helpful to come at things the other way around first. The subject of rivers in Colorado is like that.

Once, in a dry year, I walked the banks of a river where the claws of drought had dug deep. It could have been any one of the scores of Colorado rivers that rush out of the Rocky Mountains to slow in the plains or canyon country like wild horses after a stampede.

In other years—during spring and in the rain—I had seen this river crawl out of its banks and flow against stands of cottonwood trees a hundred yards from its course until each water-rubbed trunk shone like the worn handle of a good axe. In the dark, I had listened to the river as it passed full-blooded and secure in its banks, like the sound of distant thunder.

But this year, under a sky as blue as heron feathers and offering no hope of rain, there was only a narrow trail of water small enough to straddle. The air hung heavy with heat, and the river had pulled itself in, waiting for rain.

In its wake, the river left a long, ash-white high-water mark. Walking on boulders which were once deep beneath the water was like laying an outstretched hand in the track of a grizzly. Something with power had passed this way.

The thought of such a river never returning can make your throat go dry—like the parched high-water line.

Columbine, the state flower of Colorado, adorns the moist banks of spring rivers and streams. LARRY ULRICH.

Drought is real in Colorado. The state can count on only half the national average of precipitation. Colorado lies beyond the "dry line" of the 100th meridian where precipitation levels begin to trail off to the west like an echo in a box canyon. In much of the land to the west of the dry line, drought is as strong a force on the landscape as the flow of water. In each decade for the past century, since official weather records have been kept, at least one dry cycle can be demonstrated—a short but unmistakable time of drought.

The best place to begin to understand what an extended drought would mean to the rivers of Colorado—known as the Mother of Rivers because of all the rivers it gives the world—is in Kansas, or New Mexico, or Utah, or Nebraska, or any of the eighteen states across the West and Great Plains where Colorado-born rivers flow. Along rivers like the Arkansas, Platte, and Colorado are cities and towns clinging to the river like groves of cottonwoods. These cities and towns depend on the river to irrigate crops for food, to spin turbines that light the lights and move the gears of industry, to fill the taps of homes, and to make life in a semi-arid climate possible. Most of these cities and towns were founded for reasons flowing directly from the rivers—gold, furs, irrigation, transportation—and grew strong, nourished by the rivers. If the rivers suddenly disappeared in the middle of the night, these places would wake up with no reason to be there. Before long, they would simply dry up and blow away like tumbleweeds. The rivers of Colorado have that kind of power.

Colorado itself is an exposed throat to the fangs of drought. With few large natural lakes, the water runs off the mountains as if shot out of a gun. The majority of the state's flow occurs in a two-month period spanning mid-May to mid-July. It is simple. The rivers are here because the snow is here, and the snow is here because the mountains are here. The state's rivers are so tied to mountain weather that after a single snowless winter, most would run dry. Two years of drought and every river in the state would be still.

The 1,900 reservoirs which dot the map of Colorado have a combined capacity of approximately 8.85 million acre-feet, about one-half a normal year's precipitation. With strict conservation measures, these man-made sponges represent a two- or three-year supply at best. Crops, which account for 79 percent of the state's annual water usage, would be the first to die. The Dry Land Experiment Station in Cheyenne County estimates that crop failures would occur three years out of five in Colorado without irrigation.

Next, the cities would die. Places like Denver, Grand Junction, Pueblo, Colorado Springs, Julesburg, Greeley, and most others depend on rivers like cottonwood trees. Without the rivers, the same empty wind blowing across dried-up riverbeds would sound through city streets. Look no farther than the silent rock ruins of Mesa Verde along the Mancos River, or those at the Chimney Rock site on the Piedra, to understand how frail a grip life has here in the West, and how closely that life is tied to the rivers.

Cities and fields would be lost—and the rapids silenced. Canyons would grow no deeper. The landscape of Colorado, without its rivers, would go still. The rivers of Colorado have that kind of power.

But the rivers, even the one I walked along that dry day years ago, have come back to fill their banks. Each spring, the thunder of flowing water returns. Just as the droughts bring the state to its knees, the rivers bring the state to life. Every year more boaters, more anglers, and more tourists spend their time and money on Colorado's rivers. Every year the pulse of recreation in the state grows stronger, more vital.

On the rotunda of the state capitol building, in the cement heart of Denver, a stanza of verse has been carved into stone. The words, from Colorado poet laureate Thomas Hornsby Ferril, would have

The rivers of Colorado and the streams that flow into them have hundreds of gorgeous waterfalls such as this one, Whitmore Falls on Henson Creek, a tributary to the Gunnison River.
JAMES FRANK.

perhaps been more fittingly scratched with a willow branch into the mud of a river bank or etched into a blue-gray reach of driftwood in some river canyon. But even in this place their poetry remains clear. The stanza begins: "Here is a land where life is written in water."

The rivers of Colorado have scrawled the signature of life across the state's landscape, in the etchings of their deep canyons, in the splatter of green wheat fields, in the rings of willow stumps, in the huts of beavers, and in the footprints of human history. Life in Colorado is indeed written in water, and its signature is the rivers.

The Rivers of Colorado looks at these waters from both shores, at the abundance rivers bring to the lives of people in the state, but also at the cost of that abundance. *The Rivers of Colorado* looks upstream, where the story of the state and its waters has been written in courage and daring, in exploitation and greed, even now and then in folly.

The story is far from over, of course. In another, very real sense, we are looking at these rivers from midstream. As long as the snows return in winter and the fullness of rain burdens spring clouds, the story will continue. It has no end.

Once, in a shack on a roadside near a river in the southwest corner of the state, I watched a Navajo woman, her eyes the color of worked copper, weaving a tapestry of a river. With thick wool threads of silver, black, gold, and a shade of green like young willow leaves, her hands drummed a rhythm like wingbeats and on the loom appeared a river—slivers of light under a rising moon. I thought of a moon I watched once on the last wild section of the Gunnison, and of what moons might have looked like on the waters of the Blue or Platte or Big Thompson when those rivers were wild. The old woman called her tapestry "Shining Waters." And when I asked how soon it would be done, she shook her head slowly. "Soon," she said, "and never."

The story of the rivers of Colorado is like that.

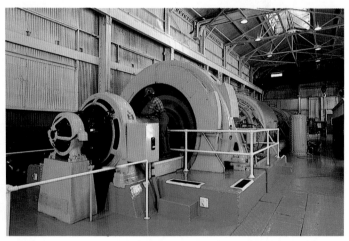

The power of rivers—what would Colorado do without it? Rivers and their tributaries provide sport for the angler, water for irrigation, fun for the floater, energy to turn turbines which, in turn, light the lights and fuel the gears of civilization.

Far left: Fishing—here in the San Juan Mountains—introduces more visitors to Colorado's rivers and streams than any other single attraction. LARRY ULRICH.

Left: Growing spinach with flood irrigation in Colorado's San Luis Valley, some of the first land in the state to feel the flow of irrigated water. DAVID MUENCH.

Top: Governor Richard Lamm (left) and Jerry Mallett floating the Yampa River. BRUCE W. HILL.

Bottom: Turbines at Shoshone Power Plant and Dam on the Colorado River. KAHNWEILER/JOHNSON.

What a watershed! In Colorado are born rivers which water states from the Gulf of Mexico to the Pacific Ocean.
Far left: Colorado's magnificent mountains capture the snow which melts into a hundred magnificent rivers.
KAHNWEILER/JOHNSON.

Above left: Snow-packed Cross Creek in the Holy Cross Wilderness, a tributary to the Eagle River. The rivers of Colorado are products of snow. With few natural lakes to buffer the dry years, a single snowless winter would still many of the state's rivers.
WARREN MARTIN HERN.

Left: A pasque flower emerges as the snow of winter recedes. LARRY ULRICH.

Above: Yellow pond lilies in the San Juan Mountains. LARRY ULRICH.

Right: Parry's primrose along one of many thousands unnamed rivulets flowing into each other to eventually form the rivers of Colorado. LARRY ULRICH.

Chapter one:

Restless earth, restless rivers

The stones look like fallen chips of a starless sky. There, on a gentle southerly bend not far from the Utah line, the Colorado River flows out of the sun-colored, young sandstones of Ruby Canyon and into the darkness of a dim and ancient past at a place known simply as Black Rocks.

The name is more descriptive than poetic, but the rocks do have their story. These 1.8 billion-year-old rocks, as fluted and polished as good gun barrel steel, are some of the oldest formations to be found along Colorado's rivers. Exposed where a river has cut a deep canyon, such as in the Black Canyon of the Gunnison or Glenwood Canyon on the Colorado, or where pressures from the restless earth have thrown them to the summits of some of the highest mountain ranges, such as the Sawatch, Gore, and Park ranges, these rocks tell the first tales of Colorado's distant past. They are the schists and gneiss and granites. Together they are called the basement complex, formed at a time when the only life on earth was boneless marine creatures. Now they are the bones of the state, the dark eyes of the past which give a glimpse into the beginnings of the long geologic story of Colorado's rivers.

The modern face of Colorado was largely sculpted by events which took place only within the last 70 million years, a tick of the clock geologically. The routes now taken by the rivers were chosen within the last few million years. But to look only at the landscape as it stands today is to jump midstream into the river of time without a glance back upstream at the events and forces which laid the foundations and set the stage.

A look upstream

Geology would be a simple matter if not for the opposing forces of erosion and uplift. Rock layers are generally laid down in neat, horizontal layers chronologically, with the younger rocks lying on top of the older formations in a package which can be read like a book, cover-to-cover. But forces pushing and pulling on the layers distort the mixture, muddying the clear views, so that it becomes harder and harder to read the stories of the rocks as they become older and older. By the time the rocks age 1.8 billion years, as in the basement complex, the language becomes almost unreadable, and blank spaces must be filled in with geologic theory.

The story of what happened during the dim past in Colorado, at a time known as the Precambrian era, is mostly a product of such theory. The Precambrian era takes in a generous slice of the earth's early history, from its beginnings 4 billion years ago to a point 570 million years before today. Rock evidence from much of this time is not found in Colorado. The oldest rocks in the state, found in the mountains in the northwest corner, date back 2.3 billion years, much younger than the age of the earth, and so the story of Colorado's beginning must be pieced together with evidence gleaned from places where the language of the rocks is more clear.

Over the globe generally, and likely in that part which would become Colorado, the Precambrian era was a time of the long, silent battles between those forces which would raise the mountains up and those which would tear them down. Several ranges of mountains probably came and went without leaving clues in Colorado. A long interval of erosion which came at the end of the Precambrian era gave the victory, temporarily, to the forces of erosion and mountains were leveled, the rocks spilling out over the land to become the floorboards of the modern continent.

As that era came to a close and the second of the four great divisions of time, the Paleozoic era, began, the land and seas began a sort of geologic dance which covered the continent alternately with shallow seas, deserts, and swamplands. These various conditions led to the laying down of some of the oldest layers of sedimentary formations found on top of the basement complex and lining Colorado's rivers—the Sawatch sandstone along the Colorado River in Glenwood Canyon, the Lodore sandstone in Lodore Canyon of the Green River, and the Hermosa shales. It was in seas of this time that the first ancestors of modern insects and fish appeared.

During the heart of the Paleozoic era, about 325 million years ago, the earth grew restless and raised a range of mountains standing very near the position of the modern Rockies. As soon as they were raised, the forces of erosion began to slowly gnaw away at their heights. In the shadows of these ancestral Rockies, the first reptilian life evolved, and in the western part of the state the Morgan sandstones, which today line the Yampa River Canyon, were deposited.

By the end of that era and the beginning of the third slice of geologic time, which was the Mesozoic era, erosion had again won out and the ancestral Rockies were leveled. The remains of the ancient mountain range can be seen today in the rocks of the Fountain Formation most prominent at the Flatirons near Boulder, the Garden of the Gods near Colorado Springs, and Red Rocks Park near Denver. Another section of the range, called the ancient Uncompahgres, had also been beaten by erosion and its sediments became the bones of

Split Mountain Gorge on the Green River in Dinosaur National Monument. Here, and in other canyons along other rivers in Colorado, the story of millions of years of geological history lies exposed for everyone to see. This canyon, as well as Gates of Lodore just upstream, was named by the Powell Expedition and helped Powell formulate his geologic theories. DAVID HISER/ASPEN.

the "Red Beds" which line the Fryingpan River today. Other formations laid down during this dawning of a new age include the sandstones—Entrada, Kayenta, and Wingate—which form the cliffs of many rivers to the west of the mountains like the lower Colorado, the Gunnison, and the Dolores.

As the Mesozoic era matured, the landscape of Colorado became webbed with the tracks of the dinosaur. Creatures like plant-eating camarsaurus and stegosaurus roamed the hot, humid lowlands which then covered the state, trying to avoid predators like the allosaurus. Some of the creatures died and were slowly covered with silt and mud from the sluggish rivers. That silt became the rocks of the Morrison Formation, the rock-type in which the rich fossil quarry of Dinosaur National Monument is found. It was also from the Morrison Formation just west of Denver that a series of fossils such as a seventy-foot brontosaurus and the three-horned triceratops were found in 1877. The town near the find was named Morrison, for the formation which has given the world the richest treasure-chest of fossilized dinosaur remains found in any type of rock.

All of these events played a part in the geology of Colorado's rivers. Yet, they took place so long ago and at such a slow pace that they are difficult to grasp. Even if they could be seen in a kind of super time-laspe photography, they would portray what would seem to human eyes a slow, uneventful period. But when the dinosaurs began to die out, giving way to the mammals and flowering plants, the stage was set for a revolution.

The revolution

In the modern sense of the term, "revolution" may seem like hardly the word for an event which occurred over a span of 40 million years. Yet, compared to the slow pace of events in the geologic time periods which preceeded the Laramide Orgeny, it must have seemed like fireworks.

At the dawning of the most recent geologic period, the Cenozoic era, the Laramide Orogeny (also known to geologists as the "revolution") began to push upward on the ancient rocks laid down so long ago in the Precambrian era, long since buried by the other younger sediments deposited on top. On and off for 40 million years the pressure continued to push and heave those ancient rocks toward the surface, buckling the earth's crust like a slowly rising fist through a layer cake. It pushed up a chain of mountains stretching from Alaska to Mexico—and thus the Rocky Mountains were born.

Most of Colorado's major ranges, housing the headwaters of the major rivers, owe their existence to the revolution. The higher ranges, like the Sawatch, Front, Park, and Gore, rose high enough to either punch through the sedimentary rocks or to expose them to the sharp claws of wind and rain erosion which quickly stripped them clean. The result, in a sense, was that the basement was in the attic. The same basement complex rocks which are exposed by the down-cutting of rivers stares down from the highest mountain peaks, pushed there by the revolution.

Colorado would become the highest state in the Union when all the uplifting was done, boasting an average elevation of 6,800 feet above the seas which so often drowned the region in the past. As a result of the revolution and a series of uplifts which followed, the lands of Colorado rose to such an extent that, of all the U.S. land above 10,000 feet, 75 percent lies within this state. Of the 67 U.S. peaks over 14,000 feet, 53 of them are in Colorado. Another 830 peaks exceed 11,000 feet.

The effects of the revolution were not limited to Colorado. The event lifted the entire five-thousand-mile-long chain of the Rockies stretching from the Brooks Range in Alaska to the Sierra Madre Oriental Range in eastern Mexico. Today, it

Ancient black rocks along the Colorado River—exposed Precambian rock—are approximately 1.8 billion years old, and among the oldest rocks exposed anywhere in Colorado. Yet geologists believe several mountain ranges were raised and erased in Colorado even before these black rocks were formed. SHERPA—SAN.

is the longest mountain barrier in the world and reaches its greatest width, over three hundred miles, at its heart in Colorado and Utah.

The Laramide Orogeny was also part of a global mountain-building episode which gave the world such great ranges as the Andes in South America, the Alps in Switzerland, and the highest of the world's peaks in the Himalayas.

The revolution may not be over. Another, more localized episode of uplift began 5 million years ago and may still be lifting the Rockies slightly. Clearly, though, the deed has been done. The Laramide Orogeny lifted Colorado to the roof of the continent, parting the waters which now fall towards two oceans and setting the stage for the rivers to become fountains of the West.

The fire

The revolution was responsible for most of the mountain ranges in Colorado, but not all. The largest range in the state—in fact the largest single range in the U.S. Rockies—had a different and fiery beginning.

Just after the start of the Laramide Orogeny, and in some ways because of it, a series of violent volcanic eruptions rocked the southern and southwestern parts of the state. Centered in what is now the San Juan Mountains, dozens of volcanic vents appeared and spewed tons of dust, smoke, ash, and lava over an area of ten thousand square miles, burying it in places more than a mile deep. Many of the tall peaks of the San Juan Range were born during this fiery period, and many of the rivers which now flow from the melting snows of these dormant volcanoes cut their paths deep into the lava-capped rocks, forming narrow and steep canyons.

The volcanic activity ceased about four thousand years ago but the fire left its mark. Not only was the largest range in the Rockies formed, but volcanic action also pushed up the ridges of

Coronet Falls on Coronet Creek, a tributary of the San Miguel River, is among dozens of spectacular waterfalls near Telluride. KAHNWEILER/JOHNSON.

Spanish Peaks at the headwaters of the Cucharas, Apishapa, and Purgatoire rivers, raised the famous Huerfano Butte which became a landmark for travelers along the Huerfano and Greenhorn rivers, and covered many of the plateaus with caps of erosion-resistant lava which have protected the White River Plateau (at the headwaters of the Yampa and White Rivers), Grand Mesa, and the other buttes and mesas which form the eerie landscape of the Plateau Country and its rivers today.

Colorado was transformed by a true revolution—rivers of melted ice and snow now follow paths once taken by rivers of molten lava and fire.

The ice

The wind, even in summer, off Arapaho Glacier in the Indian Peak Wilderness can make one shiver to the bone, a reminder of what it was like in those times when summer never came to Colorado's high country. For glaciers to form, the year-long freeze of the polar regions is not necessary—only that more snow fall in the winter than can be melted in the summer. The raising of the mountains, the blocking of the sun's rays by the smoke of volcanoes, and changes in the global climate all played a part in the formation of the icy claws which four times gouged at Colorado over the past 2 million years.

Colorado was not in the path of the great continental ice sheets which scraped down from the North, covering much of Canada and the U.S. during the Pleistocene era. Instead, some of the same conditions led to the development of smaller and more localized glacial formations in the folds of some of the higher mountain ranges like the San Juan, Sawatch, Sangre de Cristos, Park, and Gore ranges. From high up on the peaks of these mountains, glaciers reached down to a level of about eight thousand feet where the sun began to grow strong enough to halt their progress.

Because the ice never reached much below eight thousand feet, its mark is seen most clearly in the upper basins of the rivers. Valleys like that of the Eagle are piled high with thick deposits of gravel and rock gathered by the snaking ice and deposited in long mounds known as lateral moraines. Too small a river to clear its channel, the Eagle has been forced to braid itself around tons of glacial debris and weave its way downstream.

In other places, like high on the Colorado River, terminal moraines have been left—half moons of rock debris damming the ends of valleys where the glaciers suddenly stopped and began to retreat, dropping huge loads of rock carried from far upstream. Grand Lake, a large body of water on the upper Colorado, was formed by such a gathering of glacial rock.

In still other places, upper sections of rivers now flow through wide U-shaped valleys gouged out by the relentless ice. One such valley is sliced by the Big Thompson River before it slips into its deep,

Restless earth, restless rivers / 20

Above: The claws of glaciers are still dug into the landscape of Colorado—although now, they merely scratch the surface compared to ages past, when mighty glaciers gouged many of the state's river valleys out of solid rock. Remnants of these glaciers, like the St. Mary's Glacier near Idaho Springs, lay like cold, white reminders of past glory. JEFF BLACK/AMWEST.

Right: Wagon Wheel Point on the Yampa River. Sedimentary rock layers of ancient seas which once covered Colorado were pushed up by the modern Rockies during the Laramide Revolution and then exposed by the forces of erosion. Powell theorized that the Yampa existed even before such uplifts, which rose very gradually across the Yampa's path while it chewed great canyons into them. GERRY WOLFE.

The Tiger Wall on the Yampa River—one sample of the famous slickrock formations of Weber Sandstone found on the Yampa and the lower canyons of other rivers in Colorado. Here, magnesium oxide leaching has formed black streaks on the rock wall. SHERPA—SAN.

narrow, and unglaciated canyons just downstream.

The ice, however, changed more than just what it touched. When these immense ice sheets began to retreat with the warming of the climate, the meltwaters flowed downstream, raising the levels of rivers even in canyons far below where the ice had reached. Gathering debris and silt as they came, the added waters acted as a sharpened saw to help deepen downstream canyons. Many of the premier, unglaciated canyons of the West received such a helping hand with their task—Royal Gorge, Black Canyon of the Gunnison, Glenwood Canyon, and the Big Thompson canyon.

Glaciers of a smaller scale, like the Arapaho, St. Mary's, and several in Rocky Mountain National Park, still remain and send icy winds, even in summer, as a way of making us recall the touch of ice on Colorado's landscape.

Dinosaur tracks, the shadows of ancient mountain ranges, and tongues of glacial ice all seem a long way from the story of Colorado's rivers. Yet, it was these forces which set the complex and beautiful stage on which the rivers now perform. It is a three-part stage—the High Plains, the Rocky Mountains, and the Plateau Country. Because the Continental Divide wanders through the central part of the state, many of the rivers flow out across more than one of these provinces, adding to the complexity and variety of their courses.

Rivers of the plains

From the rise of a small hillock somewhere east of Denver, far enough out so that the mountains appear as just a hint of color on the horizon, it is not hard to imagine yourself standing on the floor of an ancient sea, and you are.

The plains of Colorado stretch on a bed of ocean deposits and the grains of eroded mountains from the eastern borders to the base of the foothills, bending upwards as they near the Front Range and then cresting like an ocean wave at the Flatirons near Boulder where they wash up against the bulk of the modern Rockies. The soft rock formations, in places up to six miles deep, lack major obstacles and give the rivers the run of the land. It is the classic tale of rivers choosing the path of least resistance, a situation called consequent drainage by geologists. To the rivers, it is just freedom to follow their surge to the sea, and the plains are filled with the gentle meanders of rivers with time on their hands, making the slow passage through the nation's heartland.

Despite the easy going, there are few true rivers of the plains in Colorado. Low precipitation levels along their routes provide few tributaries. Those few plains rivers which do flow—the Arikaree, Smokey Hill, and the North and South Forks of the Republican—are dusty, whimsical rivers dependent upon the scant snows of winter and an occasional summer thunderstorm.

The mountains do send two major rivers across the plains province: the Arkansas to the south and the South Platte to the northeast. Dropping quickly from their sources in the mountains, both rivers make dramatic entrances onto the plains through deep and scenic canyons.

The Arkansas, just before it reaches the plains at Canon City, cuts the famous Royal Gorge. Unlike the plains rivers which took the path of least resistance, the Arkansas established its route in younger, sedimentary rocks when the hard Precambrian rocks which now make up the walls of the Royal Gorge began to rise. With the power of dropping five thousand feet in its first 125 miles, the Arkansas River did not turn away from the rising rocks; instead it cut its way into them faster than they could rise and so maintained its course. As the walls rose around it, the river chewed its way down in a process known as antecedent drainage. The result is the dizzying heights of the twelve-hundred-foot-deep, eight-mile-long Royal Gorge.

The South Platte enters the plains through Waterton Canyon, home to both a herd of bighorn sheep and the newest undertaking of the Denver Water Board, the Strontia Springs Dam. The deep canyon is slowly losing the herd of bighorns, at least in part because of pressures from that new and controversial water project.

Once out on the plains, the two rivers continue their work, forming a pair of long, appendage-like valleys carved below the level of the surrounding plains. These river valleys are known as piedmonts. The wide loops of the meandering rivers which created these piedmonts also deposited rich soils. With the productivity of the soils, and water supplied through irrigation, both the Arkansas and the South Platte valleys have become important agricultural regions.

The piedmonts do more than nurture sugar beets and alfalfa. Geographically, they divide the eastern plains of Colorado into three sections. The northern part, above the South Platte River, forms the routes of both the Big Thompson and lower Poudre rivers, tributaries to the South Platte. The middle section, between the Arkansas and the South Platte, is similar geologically to the northern region and holds the headwaters and courses of the true plains rivers, the Arikaree, the Smokey Hill, and the two forks of the Republican.

To the south, there is a difference—lava. Eruptions from the Spanish Peaks volcanoes northwest of Trinidad buried much of the southern plains deep in ash and lava about the time the great volcanic upheavals were creating the San Juan Range to the west. Today, the rivers which radiate from the flanks of the Spanish Peaks, as well as those that flow east from the Wet Mountains and the Culebra Range, are forced to cut their courses in lava-capped formations. That means that rivers like the Apishapa, the Cuchara, the Huerfano, and the Purgatoire, all southern tributaries of the Arkansas, cut deeper and more narrow canyons than are found elsewhere on the High Plains. These canyons are the shadow-filled exceptions to the rule of gentle, meandering plains rivers sliding slowly to the sea.

Rivers of the mountains

It is a romantic notion—straddling the Continental Divide, arms outstretched, you expect that the raindrops trailing off your right hand are dripping towards the Gulf of Mexico, while those dripping off your left hand head for the Pacific Ocean. In reality, 75 percent of those drops will evaporate long before reaching the sea, and a majority of the rest will find themselves sucked up by cottonwood trees or irrigated fields, pumped into the hot tubs of Denver, or mixed with whiskey in a Vegas hotel, never getting near the smell of salty sea breezes. But perhaps it is best not to spoil the illusion.

The words "Continental Divide" conjure up a host of illusions. Perhaps the most prevalent is the idea that the Divide is a straight line of axe-sharp peaks turned upright to split each raindrop in half, a row of only the highest peaks. Here, in Colorado, that illusion comes closest to reality. Where the Divide crosses summits like those of the Indian Peaks on the Sawatch Range, it is indeed a row of beautiful, snow-capped mountains. In other places, though, it is hardly more than a bump in the road.

And it is hardly a straight line. The Divide enters Colorado from Wyoming at the tip of the Park Range, making a quick rush to the east to run through the heart of Rocky Mountain National Park and into the Indian Peaks Wilderness. There, it continues south along the Front Range and swings west again to ride the crest of the Sawatch Range. Continuing south, it makes a looping switchback into the San Juan Mountains and leaves the state at Cumbres Pass—hardly a straight line over the highest peaks. In fact, the loftiest point of all of the southern Rockies, Mount Elbert at 14,433 feet, is left standing forgotten to the east of the line.

Not to spoil the illusion entirely, the Divide does send waters to both the Gulf of Mexico and the Pacific Ocean. But perhaps a more important effect on the rivers is the Divide's influence on climate. As the dark clouds soaked with rain and snow drift towards Colorado on predominating winds from the Pacific, they confront the Divide as a formidable obstacle. Rising to the challenge, the air mass is pushed up and so cooled. Since cold mountain air can hold less moisture than warm ocean air, most of the rain or snow is squeezed out of the clouds. It falls, not on the Divide to be shared equally, but mostly on the western slope. Up to 75 percent of the state's precipitation falls on the west side of the Divide, 60 percent in the Colorado River Basin alone. By the time the clouds scrape over the mountains to the East Slope, they have been sucked dry, leaving little for the rivers there. The result is rain shadow, and it makes the eastern river canyons much different from those on the West Slope.

Rain and snow are beautiful to watch on a stormy night. They are also powerful agents of erosion. Where they fall in great amounts, like on the river canyons of the western mountains, they quickly chip away at the canyon walls. So as the river is cutting down, the rain and snow widen the canyon, making for more open and light-filled valleys like those of the upper Yampa, the Little Snake, and the green pastures of the Roaring Fork.

On the eastern side, which gets less precipitation to chew away at the walls, the main direction of the erosion is downward, caused by the river. The canyon walls stand up straighter against the elements. The result is the steep, narrow, and short canyons of the East Slope rivers. The Big Thompson, the

An oxbow lake on the Animas River. Such lakes are actually misplaced river bends where a river has sharply meandered and eventually tied itself in a knot. At a tight bend in a river, erosion is strongest at the outside of the bend where the current is fastest. The uneven rates of erosion let the river cut across the turn and take a new course, isolating the previous channel and forming an oxbow lake. SCOTT S. WARREN.

In Serpentine Bends, where the Yampa River flows through soft sandstone formations. Here, the Yampa slowly meanders around seven miles of canyons to travel a single mile—as the raven flies. DAVID HISER/ASPEN.

Poudre, and the South Platte come out of the mountains as if shot from a gun.

Geologic stories, however, are never as simple as water flowing downhill or rain crossing the Divide. Another important factor on the profile of a river is the type of rock it cuts through. Each of the three major rock types—igneous, metamorphic, and sedimentary—are represented along the courses of Colorado's rivers and each of the three has a different style in forming a canyon.

High in the mountains, pushed there by the Laramide revolution and poured there by the volcanic eruptions, are the igneous rocks. These are formations of once-molten flows hardened into granite, basalt, and associated rocks. These formations give a river little say in its course, allowing only tight, narrow canyons such as those at Gore Canyon on the upper Colorado and the narrow slots of the upper Animas River. When rockslides clog the channels with boulders, large rapids are formed since these small, high country rivers usually lack the power to erode or shove aside the hard, igneous rocks.

Metamorphic rock, like igneous rock, has been displaced by the forces of uplift and erosion and can be found both on the high peaks of the Rockies and in some of its deeper river canyons. The word "metamorphic" means changed, and rocks of this type are those which have undergone transformations due to intense heat, pressure, or exposure to natural chemical agents. This process can create the night-colored rocks at Black Rocks on the Colorado River, the quartzites which are a part of the flame-red formations which swoop up at the Gates of Lodore on the Green River, or the moon-white marble found along the Crystal River, depending on the composition of the baserock that is metamorphosed. Like igneous rocks, metamorphic rocks usually create deep and narrow canyons because of their resistance to erosion.

More subject to the artistry of water are the sedimentary formations. These are what geologists call "junk" formations, formed of various compacted sediments from shallow oceans or sand dunes or steaming swamps—limestone, sandstone, and shale. Sedimentary formations are lighter and softer and let the sun into canyons like those on the lower stretches of the Yampa, the Dolores, and the Colorado. It is in this rock that the poetic nature of water is best displayed—the meanders at Serpentine Bends on the Yampa where the river loops for seven miles to cover a single gunshot mile; Muleshoe Bend in Slickrock Canyon where the Dolores nearly ties itself in a knot; and the arches of Rattlesnake Canyon where the wind and water have whittled windows in solid rock up a sidecanyon off the Colorado. Here, the river is the artist in rock. The sedimentary rocks which give the water its poetic license are the youngest and so are most often found at the tops of canyon walls.

With the powers of uplift and erosion stirring things up, no river in the state flows through just a single rock type from headwaters to mouth. In places, the young rock of the sedimentary sits on the blackened, battered old Precambrian rock with a gap of millions of years between them missing in the rock record. At places like Glenwood Springs and the Gunnison Gorge where this occurs, it is called The Great Unconformity. In others, the formations have been tilted and faulted, mixing the layers like a deck of dropped playing cards. In all, the canyon walls and riverbanks of Colorado's rivers are most often a tapestry of rock types woven together by the hands of time to tell an epic tale of the earth's history for all with the patience to look closely.

The parks

The stories of several of the state's major rivers begin not in the high peaks but in the laps of the four great parks nestled high in the mountains. Like inverted islands, the four parks drift below the sea of high peaks which forms the Continental Divide. Surrounded by peaks, lush with water and vegetation, the parks are havens for wildlife and the headwaters of rivers.

The northernmost park was also the last to be discovered by mountain men, settlers, and prospectors. Called "the Bull Pen" by the Indians for the vast herds of buffalo and elk which wintered

The Black Canyon of the Gunnison, the oldest river canyon in Colorado. The Gunnison was already firmly established in its course, cutting through soft sedimentary rock, when it exposed the Precambrian rocks which now line the canyon walls. With no escape, the river bit slowly into the hard rock and sliced what geologists call a "superimposed" drainage pattern. WILLARD CLAY.

there, North Park was prime hunting ground for Indians and settlers alike, causing some skirmishes and classic tales of battle. In its midst, a series of small creeks flow together like the strings of some musical instrument and form the headwaters of the North Platte River.

The quiet and remoteness of Middle Park, with its sky-blue rivulets of water flowing down the slopes and coming together like the spokes of a wheel, hardly seems appropriate for the headwaters of so mighty a river as the Colorado. But here the master is born, quietly and in solitude, to begin its journey of over fourteen hundred miles through the Grand Canyon and finally to the sea.

South Park sits in the clouds. At nearly ten thousand feet, it is higher than most lands in the United States, but surrounded as it is by the Tarryall, Tenmile, Buffalo Peaks, and Mount Evans ranges, even its height seems dwarfed. From the heart of this island in the sky, the South Platte River flows first south and then turns to the north in a formation known as the "elbow of capture." It is at the apex of the swing that the South Platte, in the Tertiary period, stole the waters of a series of creeks once headed for the Arkansas River. Adding the creeks to its own flow, the South Platte used the strength to cut a series of canyons and valleys below. This kind of alluvial piracy, called stream capture, has added to the rivers of some of the best-known canyons in the West and left others forgotten and as dry as the echo from an abandoned well.

The final of the four parks is also the farthest south and the largest. San Luis Valley is tucked in between the San Juan Mountains and the Sangre de Cristo Range. Although no major rivers have their headwaters here, the Rio Grande flows into it

The ultimate in stream piracy

The dotted lines below indicate ancestral courses of present-day rivers. At left, the young Colorado has captured the Gunnison and Roaring Fork and is eroding toward the Blue, Williams Fork, and Fraser, which once comprised the headwaters of the White River.

At center, the Colorado has pirated the White's tributaries. The White has retreated toward the northwest. The Colorado has also pirated the Eagle from the Roaring Fork. The Dolores flows southwest into the San Juan.

At right, the modern Colorado has been captured by its own tributary, abandoning Unaweep Canyon. The Dolores flows north into the Colorado via the captured San Miguel.

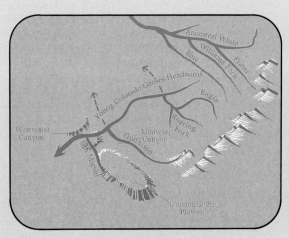 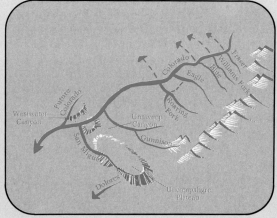 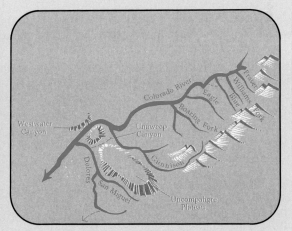

Source: The Floater's Guide to Colorado

from the West and turns south, using the course of the valley to miss the Sangres and head for the New Mexico line and deeper canyons beyond.

Almost all of the major rivers of the state have their headwaters in these mountain areas, from the high peaks or from the isolated parks. From the deep snows of the Rockies, these rivers flow out to water not only Colorado but the entire region. A total of eighteen states in the West and Great Plains have rivers flowing within their borders which began as rain or melting snow in Colorado, making it truly the Headwaters State. And it all begins at that romantically magical spot high on the Continental Divide where, with outstretched arms, it is possible to feel the faint tuggings on your fingertips of two very distant oceans.

Rivers of the plateaus

The water of a plateau river sounds like a grass fire, like bacon frying in a cast iron skillet. Churning with silt, it looks rough enough to scour off a man's whiskers in a dunking. Such rivers have come down from the mountains, out onto the plateau country. They have become different rivers.

Where once these same rivers, rivers like the Colorado, Gunnison, and Yampa, ran as clear as the mountain air, by the time they reach the heart of plateau country they've taken on enough grit to sizzle. The rivers are now plowing through sandstone, earning the "too-thick-to-drink, too-thin-to-plow" reputation of desert rivers. They flow like a gravelly voice, saying that the rivers are restless in the plateaus.

Erosion is an awesome force. It is estimated that without the compensating forces of uplift, erosion could turn the whole globe into a featureless cue ball in just 25 million years. Erosion is powerful, and its strongest punch lies in water. The muddy waters flowing in the rivers of the plateau country are telling a tale of work done upstream, of another part of a mountain range brought down, of another canyon made deeper, of a mission accomplished. Erosion is the story of the plateau province.

Plateau country could as easily go by the name "canyon country." There is as much land carved with canyons as raised into plateaus, buttes, and mesas. The uplifted land earns its name simply by providing the water which has cut the canyons below.

The largest of the trio of major uplifted sections in western Colorado is called the White River Plateau. It is a fifty-mile-wide stage in the western sky, raised by the uplift and volcanic activity which rocked the West in waves over long geologic periods. Over its northern edge spill the waters of the Yampa River, which will cut canyons like Cross Mountain, Duffy, Juniper, and Yampa on its way to the Green. Over its western edge spill the waters of the White River, which also flows to the Green River, cutting the White River Canyon along its route.

Directly south of the White River Plateau is the smaller Grand Mesa. Kept from eroding into just another patch of canyon country lowland by the lava flows centered at the White River Plateau, it owes its continued presence to the Tertiary period. Waters from Grand Mesa flow down into the basin of the Roaring Fork River as well as the many tributary streams of the Gunnison.

The longest and skinniest of Colorado's trio of plateaus is the Uncompahgre Plateau. Its folds and weathering patterns have contributed to the formations within Colorado National Monument southwest of Grand Junction and to the geologic wonders up Rattlesnake Canyon along the Colorado River.

The Uncompahgre Plateau was also the stage for one of the most graphic examples of the process of stream capture, or stream piracy, in the West. Streams generally grow headward, lengthening themselves in an upstream direction, and they may compete. A softer bed, steeper gradient, or larger quantity of water may give one river greater eroding power than another, allowing it to grow headward more rapidly and eventually penetrate the channel of the weaker.

Fifteen million years ago, the Colorado River coursed through a small, narrow gap at the northern tip of the Plateau. As the Colorado cut headward, the Plateau began to rise, slowly closing off the route of the Colorado River through the gap at its tip. A downstream tributary of the slowed Colorado gnawed its way deeper upstream, capturing the Colorado River at a point upstream from the Plateau and taking it around the northern edge of the Plateau, where it began to cut such canyons as Horsethief and Ruby in Colorado and Westwater Canyon across the Utah line.

Where the river once flowed sits an empty shell. Unaweep Canyon has the eeriness of a ruin, a great bone hollowed out and bleaching in the sun. Just two undersized creeks now flow from the canyon, splitting on a divide in its heart to run in opposite directions. Today in Unaweep (the Indian word means "parting of the waters"), there is only the wind, echoing now and again with a sound hauntingly like that from an old shell held quietly to the ear.

Geology in fast forward

The transformation of the Colorado took millions of years. The laying down of the sediments now being washed slowly away by the plateau country rivers took hundreds of millions of years. But geology is not always such a slow process. From the tangles of driftwood, as silver as worn coins, tucked high in the cliff face, and from the gaping mouths of the cutbanks, you can tell that here is a country which carries its threat of flash floods like a loaded six-gun. Flash floods are geology in fast motion.

One such flash flood took place on June 10, 1965. Before that night, the Yampa River barely rippled as it slid past the mouth of Warm Springs Canyon four miles upstream from its confluence with the Green River in Dinosaur National Monument. Major John Wesley Powell, explorer of the canyons and rivers of the plateau country, made no mention of it although he followed the Yampa upstream several miles from the Green on his 1869 expedition. It is no wonder. In 1869 there was little to say about the river at Warm Springs. He was ninety-six years too early.

The rains came hard and often during the late spring of 1965. In twenty-one days between late May and early June the canyons of Dinosaur National Monument, which often forget what rain is like, saw seventeen days of rain, hard and relentless rain. The thin soils soaked up what they could and then began to turn more water than dirt and slide off the slopes in mudflows. As the darkness fell on the night of June 10, the rains worsened and the canyon could hold no more.

A powerful wave gathered the mud, boulders, logs, and water that the canyon could no longer contain and hurled them out in a huge debris fan, damming up the river for a time at the mouth of Warm Springs Canyon. Downstream, a group of riverside campers looked on in amazement as the river slowed to a trickle even in the continuing downpour.

By early morning on June 11, the river gathered itself and topped the dam at Warm Springs, hurling itself downstream again. With a grating noise like bones rolling, boulders settled in against the current, and the rapids which would become the subject of campfire tales for years to come was born—Warm Springs Rapids.

From upstream there were no signs, no warning, of what had happened around the bend, just an uncharacteristic calm in the current like the stillness before a storm. As the two-boat commercial rafting trip approached the bend upstream from Warm Springs on the morning after the flood, two experienced Yampa river guides drifted together in midstream to discuss the changes they were beginning to sense in the river and considered pulling over for a quick scout of the bend downstream. They decided to keep drifting for this, they knew, was a quiet stretch of river.

Suddenly, they were in it. Warm Springs Rapids cannot be heard from far upstream because of the bend in the canyon walls, but it is not hard to imagine the sound that even today is like a freight train in a tunnel, ripping shouts of surprise from the throats of the boatmen and their crews as they were pulled into a rapids which hours ago was not there. It was too late to pull over, with nothing left to do but run the rapids.

It must have come hard and fast. The first boatman, straining to maneuver the heavy raft, snapped an oar, tossing him backwards and into the river. By the time the second boatman had his raft

Above: Glenwood Canyon on the Colorado River. Geological events can, at times, happen rapidly. When rivers swell from abnormal amounts of rainfall or spring runoff—as in the floods which swept the Colorado River in the spring of 1983—the process of erosion can go into fast forward. DANN COFFEY/THE STOCK BROKER.

Left: Destruction along the Roaring River when Lawn Lake Dam in Rocky Mountain National Park, an earthen structure, gave way one summer morning in 1982 and turned downtown Estes Park into a raging river, leaving three people dead and $30 million in property damage. An early morning delivery driver discovered the flood and gave warning, saving countless lives. JOE ARNOLD JR.

safely in an eddy below, it was too late. The untended raft had somehow made it through with the passengers safe but the boatman was nowhere to be found. As they spent the following hours scrambling over the rough rocks along the shore, it became apparent that the second boatman was lost. They had no choice but to give up the search. Over two weeks later, the body of Les Oldham was pulled from the river at Island Park, fifteen miles downstream.

Warm Springs remains the most respected rapids in Dinosaur National Monument, but today hundreds of boaters each year run it successfully. The rapids has settled and stabilized, becoming less severe; and with proper precautions, good equipment, scouting from shore, and a dose of luck, the rapids can be run safely at most water levels. Still, the roar is like a gunshot in a box canyon and the waves of the river's newest rapids seem as big as the canyon walls whenever you round the bend. And no matter how many times you've safely paddled it, each time the boat pulls up to that hundred-foot fan of truck-sized boulders left by the Warm Springs flood, there is that quick shiver, like a cold and rainy wind that passes—for what happened here and for the fear in the throats of the first boatmen to round that bend in the Yampa River, becoming the first to stare into the legend.

The geology of Colorado rivers is a collage of events, from the dramatic to the commonplace, from the rising of mountains to the tint of sandstone in a desert river. Each passing era and each puff of wind has left its thumbprint on the rivers of the state. Some left it boldly across the maps in a tangle of topographic lines. Some wrote it softly in the dust. Each has played a part.

Still, to speak as though the story were over would be a mistake. It is not, and the click of pebbles down a cliff face, heard as faintly as distant bells, is just the river's way of beginning yet another chapter.

Warm Springs Rapids on the Yampa River, near the Green River confluence. Before a 1965 flash flood in Warm Springs Creek, this rapids didn't exist. In 1869, Powell found only a riffle here. Now, it is the largest and most notorious rapids in Dinosaur National Monument, backing up the river for six miles behind its natural dam of tons of rock debris. TOM BECK.

The arches of Rattlesnake Canyon

There is an inherent artistry in water and wind, and its work can be seen in the walls of a little-known side canyon off the Colorado River, just downstream from Grand Junction, at a place called Rattlesnake. Situated in the same rock type as the famous Arches National Park just a hundred air miles west, the same forces of wind and water and time have been at work on the Entrada sandstone of Rattlesnake Canyon. The result is a treasure-chest of geologic formations.

An arch is the product of unveven erosion, a design of wind and rain. As the slightly acidic water of raindrops and snowflakes seeps into tiny cracks in the rock, it slowly breaks down the binding material and erodes away the softer rock first, leaving the harder layers to stand longer. The process is speeded along by water freezing and expanding in the cracks, a powerful erosive agent. Entrada sandstone, like the walls of Rattlesnake Canyon, is particularly susceptible to this uneven erosion because of its easily dissolved cementing material. As the soft rock is carried away towards the river, the hard rock stands and slowly the rock is chewed away, letting the blue sky shine through: an arch is formed.

Thirteen arches have been identified in Rattlesnake Canyon, the greatest concentration found anywhere outside of Arches National Park. There are arches a coyote would have to crawl through, others which would scrape the tallest branches of a cottonwood, and still others as large as houses (or more appropriately, cathedrals). There is even a double arch.

Also in Rattlesnake Canyon, there are rock spires hundreds of feet high, huge boulders that balance on edges of cliffs, windows in the rocks, amphitheaters which seem to echo with the wind, and other geologic treasures. The one thing missing is protection for these valuable formations. Unlike Arches National Park, Rattlesnake Canyon is currently unprotected. It has been considered for an expansion of nearby Colorado National Monument and as an addition to the National Wilderness Preservation System, but as of this writing, it stands alone. It is a place where the forces of nature have touched the earth with beauty and like all great art, it deserves to be admired, wondered at, and protected.

The arches of Rattlesnake Canyon on the Colorado River, one of many hidden treasures along the rivers of Colorado. JEFF RENNICKE.

Chapter two:

Always looking around the next bend

It is late, and with no moon shining above the canyon walls, our camp is dark with just what soft, thin light the stars can give. From the fire, a chip of driftwood in the ash sends up a string of pale, gray smoke, braiding as it rises like the hair of an old Indian woman.

In the darkness at the river, someone is securing our rafts and kayaks for the night, pulling them snug against the current. The rap of a slamming riverbox ricochets an echo off the canyon wall as sharp as a spark, then again more distant, once more and is gone. Afterward, the camp goes quiet for the night except for the long, slow murmurings of the river counting its stones over and over in the dark. One hundred and sixteen years earlier, in 1869, the men of the Powell Expedition camped near here, where the waters of the Yampa and Green Rivers tumble together under the cliffs of Echo Park in Dinosaur National Monument.

Filling in blank spots on the map

Maj. John Wesley Powell wrote in his journal that his expedition camped in the area for three days, resting, taking measurements from the sky, and playing games with the "magic music" of the echoes which his men claimed gave back as many as "ten or twelve repetitions" of their shouts.

But the Powell Expedition would be remembered for something more substantial than an echo. They would be the first party to raft the Colorado River through the Grand Canyon, filling in one of the last remaining blank spots on maps of the western United States. The epic journey would win the Major and his crew a permanent mooring in the history of river exploration.

Passing through Lodore Canyon on the Green River, Powell would also become the most well-known player in the story of exploration of Colorado's rivers. It is a story made up of both the famous and the faceless, some that history remembers kindly and others it has long since forgotten. Men like Powell, Zebulon Pike, John Fremont, and Stephen Long are all remembered in the names of mountain peaks in the West, and Kit Carson's name is chiseled into stone at a cliff face on a tributary creek of the Arkansas River. Others are less honored, like one forty-five-year old housepainter, A. J. "Spike" Breitkruetz from a small town near the Colorado River headwaters. Breitkruetz set out in May of 1940 with just his twelve-foot homemade raft and his "lucky arrowhead ring" hoping to become the first one to raft the mighty river from its source, only to lose everything but his life and his lucky ring in one of the first rapids. He quietly returned to the less-glamorous but decidedly safer pursuit of painting houses.

All who came carried dreams—of discovering "lost" cities of legend, blazing a trail through the wilds, plundering endless riches of gold or fur, adding to the depths of scientific knowledge, or simply being first. The dream continues today as advances in equipment, safety techniques, and boating skill make it possible for the new explorers, the talented river runners, to challenge rapids and canyons which have not been safely boated by rivermen before. Sometimes, the bend is in the canyon walls or in the course of the river; sometimes it is in the human mind or our technology. But all of the explorers of Colorado's rivers have shared at least this one thing: a desire to move ahead, if only for a look around the bend.

For Gold and God

In the year 1528, a sailing vessel belonging to a Spanish exploring fleet was blown off its course and wrecked somewhere on the Gulf coast. As the survivors gathered themselves on shore and looked inland to the thousand miles of heat, cold, thirst, hunger, and hostile Indians between them and the nearest Spanish settlement, they could not know they were determining the futures of some Colorado rivers.

Eight years later, the weather-beaten and wild-looking men of that Spanish ship crawled into San Miguel de Culiacan, an early Spanish outpost, bab-

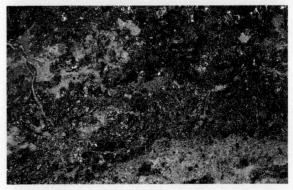

Above: The beaver, a major reason the white man came to Colorado. JIM BRANDENBURG/DRT.

Bottom: Gold, the other major incentive for the exploration of Colorado.
BRIAN PARKER/TOM STACK & ASSOCIATES.

Right: The first dams across the rivers of Colorado were beaver dams like this one on Elk Creek in the Weminuche Wilderness. LARRY ULRICH.

bling of a place where the trees were hung with golden bells and cities of gold shimmered in the endless sun. Soon, the search for the fabled Seven Cities of Cibola was on, a search that would bring some of the earliest known explorations of Colorado's rivers.

Many of the early searches for Cibola went unrecorded or have been forgotten with time. Some remain only in bits and pieces, a name on a butte and a river we know as the Huerfano, called "The Orphan" by a lonely group of Spaniards. Some remain only in legend, like the tale of the mysterious disappearance of a Spaniard expedition along the banks of a dusty river in southeastern Colorado, leaving behind only a few pieces of rusted armor and a sorrowful name for the river, El Rio de las Animas Perididas en Purgatorio, "The River of the Souls Lost in Purgatory." Today, all that is left of the tale is a name—Purgatoire River.

Don Juan Maria de Rivera came looking for gold and between 1761 and 1765 became one of the first Europeans to lay eyes on rivers like the Animas, Dolores, and Gunnison. To leave his mark on something more substantial than legend, Rivera carved a cross into a cottonwood tree on the banks of the Uncompahgre River where it meets the Gunnison. The inscription would be seen as a sign of hope by the next Spanish party to explore the area, a party which came to the rivers of Colorado for a very different reason.

Fathers Silvestre Velez de Escalante and Antanasio Dominguez left Sante Fe, New Mexico, on July 29, 1776, looking for God, not gold. It was their mission to blaze a trail which would offer easy passage to the new missions of Monterey, California, and to bring the word of God to the peoples they met along the way. Like almost all early expeditions, they were guided by the course of the rivers, following the Rio Chama north into Colorado, and there turning west towards the lower reaches of a river the Navajo called Powhuska, "the Mad River." Overlooking the Indian name, the maps drawn by the Escalante-Dominguez expedition claimed it as the San Juan, just one of dozens of rivers, creeks, and other geographic features which today wear a name given in this early expedition.

From the San Juan, they continued west, crossing and giving names to such rivers as the Piedra, La Plata, and Dolores. For eleven days they followed the waters of the Dolores and then turned north to a river they called El Rio de San Francisco, one of the few rivers in the state which has kept its original Indian name, the Uncompahgre River, a Ute word for "hotwater springs." There, where the Uncompahgre meets the Gunnison, the Fathers saw a cross carved into a tree and took it as a sign of good fortune. The party had placed themselves squarely in the hands of their God, refusing to carry weapons for protection even into known territories of hostile tribes. The sight of the cross reaffirmed them in their mission.

It did not, however, bring them much luck. Continuing up the North Fork of the Gunnison they skirted the Black Canyon of the Gunnison and continued north, fording the Colorado, White, and Green rivers while making their way out of Colorado. Just into Utah, their luck and the weather began to close in on them. Winter snows had worked their way down from the surrounding

A trail of names

Maps can be deceiving. At first glance, a map of Colorado rivers shows only the barest geographic details. But a closer look can trace the path of exploration. As the early explorers tracked the rivers of the state, they left behind more than footprints. They left a trail of names.

The Spanish were among the first Europeans to wander along the state's rivers. Much of their exploration was done along rivers in the southern part of the state, where rivers' names today ring with a Spanish accent—Rio Grande, San Juan, Piedra, Animas, Los Pinos, Dolores, San Miguel, Conejos, Huerfano, Colorado, Mancos, and Rio Chama.

French trappers and prospectors were some of the first to trace the rivers of the north and eastern part of the state, in search of gold and furs. There, the names of rivers have a French accent—like the North and South Platte (*Riviere de Plat* in French), Cache la Poudre, Laramie, Blue, and Canadian.

The rugged west-central mountains of the state were the stomping grounds of the mountain men. More often than other parties, the mountain men respected and used the original Indian names for rivers, like *Seedskeedee* for the Green and *Nahunkahrea* for the Blue; but other Colorado rivers bear names given them by the men of the mountains. Mountain men names, like their whiskey, were strong and got to the point—Elk, Fryingpan, Crystal, Roaring Fork, and East.

Only a handful of rivers in the state retain their native Indian names—the Apishapa, Yampa, Uncompahgre, Arikaree, and Arkansas. Others have been renamed in honor of explorers—the Williams Fork, named for mountain guide Bill Williams; the Fraser, which was named in honor of an officer in the expedition of Captain Bethoud in 1860; and the Gunnison, named after Captain John W. Gunnison, who led a party to its banks in 1853.

peaks and forced them south towards less inclement weather. Had a course to the west been held, Dominguez and Escalante would have found easy going to the Pacific and completion of their mission. Instead, they found starvation and cold. In January of 1777, after having been forced to eat many of their pack animals to survive, the party straggled back into Spanish settlements.

Although they did not find a route to the California missions, they did blaze part of what became known as the Old Spanish trail and gave names to many of Colorado's rivers.

In search of a ghost river

The rivers hadn't changed. They looked the same. And they still flowed downhill.

But something had changed. By the time the next major expedition came to the rivers of Colorado, those rivers were flowing on American soil. The Louisiana Purchase of 1803 had given the country its heart—827,192 square miles of wilderness stretching from the Mississippi River to the land west of the Rockies, an unexplored paradise as large as western Europe. What laid in the lonely heart of the young country even President Jefferson, who secured the lands, did not know. So two expeditions were dispatched. Meriwether Lewis and William Clark were sent into the northern reaches of the new territory, and to the south, along the rivers of a territory which would become known as Colorado, came young Zebulon Montgomery Pike.

Pike and his party of twenty-two men left St. Louis on July 15, 1806, following a trail blazed by earlier Spanish expeditions which paralleled the Arkansas River. The military orders under which the party marched seemed clear enough: "Ascertain the direction, extent and navigation of the Arkansas and Red Rivers."

Even in the early 1800s, the Arkansas was a well-known river in its lower reaches. It was the Red River which would haunt Pike and elude him as if it flowed with quicksilver.

On November 13, with winter already sharpening the air, Pike entered Colorado near the present-day town of Holly. Not far upstream, near where today stands the $17 million John A. Martin Dam, the horizon took on a more blue tint, the color of stormclouds, and Pike saw for the first time the peak which would wear his name. "At two o'clock in the afternoon," he wrote in his journal, "I thought I could distinguish a mountain to our right, which appeared as a small blue cloud." The "cloud" grew into a peak 14,110 feet in the sky as Pike drew closer, following the curves of the Arkansas River.

From the description in his journals, Pike set camp at what would seem to be the confluence of the St. Charles River. In his eyes, two of his goals were already in sight. "The river," he wrote, "appeared to be dividing itself into many smaller branches and of course must be near its extreme source..." and the mountain which he hoped to climb for a look at the countryside was surely "one day's march" from camp. On both counts, Pike was wrong.

Three days later, standing in three feet of snow with "no socks," wearing only the light cotton army uniform of summer while the thermometer read minus four degrees fahrenheit, Pike thought the peak still seemed more an illusion than a mountain. But reality was biting down with cold, hard teeth. The base of the mountain was still twelve miles distant and, from the top of the Cheyenne Ridge where Pike stood, it seemed to him that "no man could have ascended to its pinical." He turned away from the shining mountain and began the long march back to camp, eating only "one partridge and a piece of deer's ribs the ravens had left us."

The effort of the aborted climb cost Pike dearly, in spirit and in body. Precious time had been lost

The lure of gold drew more than 100,000 people to the rivers of Colorado during the rush of 1859. And even today, many of the state's rivers and tributaries give up "color" to weekend prospectors. SHERPA—SAN.

and winter dug its claws even deeper. As the camp at the St. Charles River was broken, a snowstorm whipped the faces of the men, hiking into the wind following the path of the Arkansas.

Through the blizzard, Pike made an unexpected discovery—the Royal Gorge. Although it did not make up for his failure to climb the peak, the discovery of the gorge was a true geographic accomplishment, the importance of which Pike himself may not have realized. Marking the gorge on his maps, Pike turned north to avoid the darkness of the canyon and towards another unexpected and important discovery.

Like the lower Arkansas, the Platte was a well-known river where it etched into the long, flat back of the prairie. Yet its headwaters remained a mystery. On December 13, Pike wrote, he and his men came upon "a river 40 yards wide, frozen over which...I found to run northeast." Despite all the errors on the early maps, Pike knew the Red

River, if he ever found it, would flow southeast. There could be only one logical explanation for this wide river flowing northeast—the headwaters of the South Platte, and at a point much farther west than the map-makers had placed it.

It was another significant geographic discovery for Pike, but somewhere out there lay his goal, the one stroke of the map-maker's pen which he longed for—the Red River, the ghost river.

Turning west, he began his dance with the ghost river. Five days after his discovery of the South Platte, he came upon another river which "some of our lads supposed to be...(the) Red River." Turning upstream, they traced the mystery river to its headwaters near the Sawatch Mountains and then began climbing back down its course.

Then winter took a hard bite. The frozen river made a difficult trail, battering and injuring the pack animals as they struggled to keep footing. Three were shot and left behind. Sleds were constructed for the gear the animals could not carry. The walls of the canyons—first Brown's Canyon and then the deeper Arkansas River Canyon —reached up to block out the light of day.

From the depths of despair, Pike left the group to climb a ridge to see which route offered the best path. As he reached the height of land and the countryside unfolded before his eyes, he discovered, "with great mortification," that he had led the men in a wide circle and the river they were on was not the Red but the Arkansas, upstream from where the Royal Gorge had turned them north a month of cold campsites earlier.

It would have been easy for Pike to turn back from his mission. He had, after all, discovered the sources of the Arkansas and the South Platte, carved a deep and wonderous canyon into the maps, and come eye-to-eye with Pike's Peak. The route was open to him, down the Arkansas, around the string of peaks to the east and across the plains —just the way he came. Instead, Pike looked away from the easy route and his eyes fell on the formidable crags of the Sangre de Cristo Range and the dream of what might lay beyond.

In the heart of a Colorado winter, Pike successfully led his men over the Sangres on a route which took them through Medano Pass and brought them to two more discoveries. One was the great waves of sand tucked up against the southwestern foot of the mountains which, as Pike described them, looked "exactly (like)...the sea in a storm except as to color." The dunes he saw are now part of the Great Sand Dunes National Monument.

From the top of the pass, he made the second discovery—a river shining on the valley floor. It was a river, Pike wrote, "hailed with fervency as the waters of the Red." But once again, like a mirage drifting across the sand dunes, the ghost of the Red River had come to haunt Pike. This was the Rio Grande River.

In the weeks that followed, still convinced he had found the elusive Red River, Pike erected a small fort just west of the Conejos River, sketching it on maps as the west branch of the Red. The fort was mistakenly constructed on Spanish soil, Pike's final error. On February 26, 1807, a unit of the Spanish army rode up and arrested Pike, taking him into custody and bringing his expedition to an unceremonious end.

Pike was brought to Sante Fe and later to Mexico before his release could be secured. In an ironic twist of fate, the route Pike followed upon his release took him into the Louisiana Territory, along the river which had so frustrated him on his journey into Colorado, the Red River. Pike had been told the Red River flowed from the Rockies, but in reality it was hundreds of miles to the southeast in northern Texas.

Others would follow Pike—Long, who came in 1820 to fill in what blanks had been left by Pike's arrest; Fremont, in four expeditions between 1843 and 1852, and Gunnison in 1853, both searching the maze of mountain passes for a railroad route; and Hayden on his great surveys of the West with photographer William Henry Jackson, through whose lenses the world would first see the scenic wonders of the region.

All of them followed, at least in part, the path of Pike. Some have called Pike's expedition a failure, and his arrest was a scandal which haunted him for the rest of his days. But his accomplishments stand as proud as his mountain, and he is hailed as a hero in the story of Colorado exploration. In his search for a river that was not there, he gave the world some of its first glimpses of the power and beauty of those rivers that were.

The dark heart of the Royal Gorge on the Arkansas River turned back both the men of the Pike Expedition who discovered it and the Long Expedition who followed. Today, the Royal Gorge attracts tens of thousands of tourists each year to the world's highest suspension bridge dangling 1,053 feet above the river.
PAUL LOGSDON.

Willow sticks and Willowas

The mountain men were as straightforward as a wind from the north—strong, practical, and bent on doing things to get them done. Rivers meant one of three things: creeks for beaver; color in the bottom of a gold pan; or a watery obstacle that had to be forded. Beaver pelts and gold got the most attention. But the way the mountain men solved the problem of getting to the other side of a river, or downstream a spell, led to important advances in the exploration of rivers. They solved the problem in the Indian fashion, with willow sticks and buffalo hides.

Bull boats were functional if not beautiful. When a river crossing was necessary and there happened to be a stand of willows nearby and a few buffalo or elk hides handy, a bull boat was likely to be the result. And once the crossing was completed, these hastily-constructed crafts could be discarded.

Construction of bull boats was as individual as the maker, but all were a variation of the standard theme of a woven basket of willow or young cottonwood branches covered with the hides of elk or buffalo (hair side in), sewn together with leather thongs, and waterproofed with a mixture of tallow and ashes or with pitch made from pine tar. They floated like autumn leaves even with a load, the hides staining dark as bark against the water. Some were just big enough for one paddler and gear, but others ranged up to thirty feet long and could carry, according to one early bull boatman, "three men, sixty traps, five hundred beaver pelts, guns, ammunition, and some miscellaneous goods."

The advantages were many—ease of construction (Jim Bridger, an expert bull boater, could fashion a "very respectable" craft in a matter of hours); shallow draft under heavy loads; light enough to portage when necessary; and, on a rainy night, the smaller boats could be tipped upside-down over the chimney holes of teepee or shelters to keep the rain from ruining a night's sleep by the fire.

Disadvantages were many as well, and became obvious with extended use. The circular shape made them difficult to control in strong current; the water-proofing scraped off on the sandbars or was chewed off by the dogs; if left in the water too long the skins became soft; if left in the sun too long the seams split; as the boats got older they took on what is best described by a bull boatman on the Gallatin River of Montana as an odor "far from balmy"; and, because of the natural materials, the wolves which often trailed the parties found the bull boat a particularly fancy feast.

It was, for the most part, a flatwater boat. Still, Lewis and Clark, who augmented their flotilla with at least two bull boats, reported that they offered "perfect security" even through "the most difficult choals and rapids." Others, like Thomas "Broken Hand" Fitzpatrick, who lost two bull boats fully loaded with a season's pelts to the Platte, would likely argue against the craft's prowess in whitewater.

For all its faults, the bull boat won a place in Western history and the list of those who struggled with them in the currents reads like a roll call of legends. Besides the Lewis and Clark Expedition and Fitzpatrick, Jim Bridger, Toussaint Charbonneau, Kit Carson, Colonel Fremont, Robert Stuart, and Jedadiah Smith all built bull boats. Indian tribes like the Blackfoot, Assiniboine, Cree, Cheyenne, and Pawnee also put the craft to use. And bull boats were a common sight on rivers across the West such as the Yellowstone, Platte, Gila, Gallatin, Cimarron, and Missouri.

In Colorado, the bull boat was most often used where the rivers settled after spilling down from the mountainsides, on the rivers of the plains and plateau country. Places like Bent's Fort on the Arkansas, Fort Robidoux on the lower Gunnison, and the trading posts on the South Platte were all destinations of parties traveling in the tubs of willow sticks and buffalo hides.

Two early bull boaters stand out in Colorado history. The first was a man described by Jack Sumner of the Powell Expedition as "gentle as a child when used rightly, a wounded grizzly when provoked." James Baker, who claimed to have visited every valley in the Rockies, was "a handy man with his fists," according to Sumner, and

Easily constructed, efficient, and often disposable, the bull boat proved perfectly suited to the needs of early mountain men who copied it from Indians. With materials at hand, a trapper could build a one-man bull boat to cross a river, or a larger one to carry heavy loads of furs as well, and expeditions constructed bull boats as large as thirty feet long to ferry equipment.

must also have been a handy man with a paddle. Although Nathanial Galloway is often credited with the first descent of the Yampa River in 1909, Sumner claimed, in an account of the Powell journey given to Robert Brewster Stanton in 1907, that Baker and a companion "went in bull-boats down the Bear River Canyon to the Green and down the Green to Kelly's Hole." In Baker's day, and even by some old-timers today, the Yampa was known as the Bear River and Kelly's Hole is now called Island Park, a point ten miles downstream from the confluence of the Yampa (Bear) and the Green. No date is given by Sumner; but since Baker died in 1898, his exploration of the Yampa River in bull boats must have pre-dated Galloway's by at least a dozen years.

Colorado's most famous bull boatman pushed off into the current of the Green River in the spring of 1825. Three years earlier, William Ashley—the man who was to become known as the "king of the fur traders"—ran an ad in the March 20, 1822 edition of the *Missouri Republican* calling for "Enterprising Young Men" to join him on a westward adventure. The response brought Ashley a group of men whose lives would carve the face of Western history. There was Kit Carson, the guide of the West; Jim Bridger who, among other things, discovered the Great Salt Lake; Bill Williams, who led the fourth Fremont Expedition and left his name on a high tributary of the Colorado River; John Colter of "Colter's Hell," which would become better known as Yellowstone National Park; William Sublette, blazer of the Oregon Trail; Hugh Glass who, after a grizzly attack and being left for dead by his partners, would crawl two hundred miles for revenge against those who abandoned him; and Colorado's own mountain legend, James Beckwourth, who lived a double life as a storekeeper in Denver and an honorary Crow chief.

With this legion of legends, Ashley set out up the Missouri to the South Platte, splitting his group to spread out and trap any creek over the entire western Rockies where the air smelled like beaver. Ashley himself and the small party he led came up the Platte to the Cache la Poudre and crossed over the mountains heading west. After the best part of his pack train was stolen by the Crow Indians, leaving only the men to carry the gear on their backs, they struck "a beautiful river running south" on April 18, 1825. Without horses, the river seemed to offer the easiest route and so, promising to mark a post downstream for a rendezvous sometime that fall, he constructed one large bull boat and pushed off down the Green.

The story of what happened after the bull boat rounded a bend out of sight is difficult to piece together. Ashley's account is limited to a few letters, and the facts have been woven into so many fireside tales of mountain men like Jim Beckwourth, who valued their story-telling just behind their shooting piece and their top-knot, that the yarn undoubtedly has been spun a long way from the truth. Beckwourth told of "awesome sucks" and his stories were published in an 1865 biography that overshadowed the less-inspired and probably more truthful account by Ashley himself.

According to Ashley, the trip met with no formidable obstacles in the river's upper reaches. They drifted down, marked a spot at the mouth of Henry's Fork Creek for the upcoming rendezvous

Saved by the seat of his pants

The territory Powell and his men entered on their 1869 journey was as unknown and uncharted as any remaining in the West. Hostile Indians, waterfalls, wild animals, rattlesnakes, and a host of other dangers stood between them and success in their venture. As the burning of their campsite in Lodore Canyon illustrated, not every danger took the form of rocks and rapids in the river. In an attempt to climb Steamboat Rock at the confluence of the Green and Yampa rivers in northwestern Colorado to pin-point their location, Powell himself (who had only one arm) nearly fell to his death. His journals recorded the incident like this:

'Here, by making a spring, I gain a foothold in a little crevice, and grasp an angle of the rock overhead. I find I can get up no farther and cannot step back, for I dare not let go with my hand and cannot reach a foothold below without. I call to Bradley for help. He finds a way by which he can get to the top of the rock over my head, but cannot reach me. Then he looks around for some stick or limb of a tree, but finds none.... The moment is critical. Standing on my toes, my muscles begin to tremble. It is sixty or eighty feet to the foot of the precipice. If I lose my hold I shall fall to the bottom and then perhaps roll over the bench and tumble still farther down the cliff. At this instant, it occurs to Bradley to take off his drawers, which he does, and swings them down to me. I hug close to the rock, let go with my hand, seize the dangling legs, and with his assistance am enabled to gain the top."

Of all the dangers overcome by the most daring expedition in Western river-running history, of all the rapids run and all the unknown canyons explored, the expedition came near to a premature end on a cliff where the Green and Yampa Rivers meet. An expedition that would make history was saved by quick-thinking and a sturdy pair of drawers.

with his trappers, and continued downstream towards Red Canyon.

It was there, in Red Canyon, that the first rapids were confronted. It was too large to run in bull boats and so the gear was unloaded and the whole mess portaged around the bad water. Before getting back into the boat, Ashley took the time to paint an inscription on a flat rock at the rapids head: Ashley, 1825. The mark, seen by a long string of explorers who followed—Manley in 1849, Powell in 1869, and even Emory and Ellsworth Kolb on a photograph journey eighty-six years later—is now, like the rapids that it overlooked, drowned beneath the impounded waters of Flaming Gorge Reservoir.

Calm water greeted them for a time below Red Canyon and through Brown's Park. But in the distance ahead they saw a sweep of canyon walls curling up from the river like a pair of hands reaching for prayer. At the Gates of Lodore, even today, the river flows most of the day in darkness and the air in the shadows wears the chill of cold steel. It was not a feeling lost on Ashley, who wrote in one letter: "We proceeded down the river about two miles, where it again enters between two mountains and affording a channel even more contracted than before. As we passed along between these massive walls, which in great degree excluded us from the rays of heaven...I was forcibly struck with the gloom which spread over the countenances of my men; they seemed to anticipate (and not too far distant too) a dreadful termination of our voyage...for things around us had an truly awful appearance."

No such "dreadful termination" awaited them. The rapids which spill over the heart of Lodore Canyon forced them to make many long and difficult portages, and the shadows kept the days in gloom. But they did not come to the disaster about which Powell speculated when he found an abandoned boat and gear below the rapids he named Disaster Falls, out of respect for the troubles he guessed Ashley had there. Powell had much more difficulty forty-four years later in his specially designed boats than Ashley had with his bull boats.

Ashley floated deep into Desolation Canyon before striking a trail away from the river. He never made much of his journey, treating it only as a necessary danger in pursuit of his fortune, but the trip stands both as a tribute to Ashley's good boating skills, which got them safely through a canyon which would bring later expeditions to their knees, and to those thin-shelled vessels born of willow sticks and buffalo hides known to the men of the mountains as bull boats.

A chapter of disaster and toil

It had been a long, tough week. Writing in his trip log for June 17, 1869, while camped at the confluence of the Green and Yampa rivers in northwestern Colorado, Major John Wesley Powell looked back over the past ten days and the deep canyon he and his men had been spewed out of and called the whole ordeal "a chapter of disasters and toils." He was not without good reasons.

In those ten days the party, on the most historic river trip in western history, had managed to put only twenty miles beneath their boats, and most of that by carrying them around rapids. One boat, the No Name, had been lost, or more poetically as Powell put it, first "broken quite in two" and then "dashed to pieces" on the rocks when the oarsmen failed to make the pull-out above what would become known as Disaster Falls. Food, clothing and scientific instruments were lost but the salvation of a "three gallon keg of whiskey" kept the incident from being a total loss. Another boat, the Maid of the Canyon, was very nearly lost when it broke free of the ropes being used to lower it over a rapids. Only a lucky current which pushed the wayward boat into an eddy where it could be recovered kept the trip from ending almost before it began. Then the cooking gear, and almost the

The Green River as it leaves Colorado cuts a deep canyon which still carries the name given it by the Powell Expedition—Whirlpool Canyon.
STEWART M. GREEN.

cook, was lost to the river when sparks from the campfire set the whole streamside ablaze.

By the time Powell sat by the light of his campfire, listening to the voices of the men ricocheting off the canyon walls and writing in his journal, he was only twenty-four days into what would be more than a three-month trip. One man, Frank Goodman, had made up his mind to leave the expedition at the first opportunity and, from what sketchy information Powell had gathered about the river, he knew the most dangerous rapids and remote canyons were still downstream. For three days his party camped at Echo Park, fixing broken gear, biding their time, worrying, and gathering their courage like so much wind-scattered seed.

As history has recorded, Powell and his men did

Prospectors and fur traders followed the rivers up into the mountains; adventurers floated the rivers downstream to explore new territory. Today, travelers go both ways with ease. Trains and cars, carrying today's explorers, take advantage of natural grades formed by the rivers. Here in Glenwood Canyon, the Colorado River has attracted both railroad tracks and a two-lane highway which is soon to be widened.
KAHNWEILER/JOHNSON.

summon enough courage to round the bend and go on. By August 29, 1869, the walls of the Grand Canyon hundreds of miles downstream were falling away at a place called Grand Wash; the journey was over. The last great blank spot on the maps of the West had been etched in. The last of the great western adventures came to a close. At that final campsite beneath the canyon walls, Powell wrote: "The river rolls by us in silent majesty; the quiet of camp is sweet; our joy is almost ecstasy. We sit till long after midnight talking of the Grand Canyon, talking of home..."

It is a story which belongs mostly to the Grand Canyon and those canyons downstream from Colorado where Powell followed no one's trail. Only thirty-nine miles of the journey fell within Colorado, on a section of canyon which had been run at least twice before Powell. Yet the "chapter of disaster and toils" of Lodore Canyon is an important one to the Powell story, and his successes and failures there changed not only the story line of his own voyage but also the course of river exploration everywhere in the West.

Lodore, with its many rapids, was the first real test for Powell's boats. They flunked. The boats were built by an outfit in Chicago to specifications outlined by head boatman Jack Sumner. Three of the four were twenty-one feet long, six feet wide, twenty-six inches deep, and sported water-tight compartments fore and aft. The other was the pilot boat, the Emma Dean, and was sixteen feet long, four feet wide, and twenty inches deep. All were made of oak and double ribbed with reinforced bow and stern plates to withstand collisions with rocks.

And they did hit rocks. From the moment they felt the pull of the Green River's current, the drawbacks of the design became clear. Overloaded, too heavy, sluggish in the fast water, and with inexperienced hands at the oars, the boats continually hung up even on sandbars in quiet water. Andrew Hall, a young bull-whacker who had spent a good deal more time in the saddle than at the oars, claimed that the Maid of the Canyon would "neither gee, nor haw, nor whoa worth a damn." The boat, in his words, just "wasn't broke at all!"

Powell himself was ill-prepared for what the river would ask of him. As a part of a research study conducted on freshwater snails, Powell had boated on stretches of the Mississippi, Ohio, Illinois, and Des Moines rivers back East. But flatwater and snails did little to hone his skills to the level necessary on a river like the Green. As river historian Otis "Dock" Marston put it, "Major Powell was an able geologist, ethnologist, politician and philosopher, but he was not the most skilled boating leader the river has seen, by quite a long way." The combination of inadequate boats and the less-than-shining leadership did not bode well for what lay downstream where the river growled like a distant roll of thunder.

Powell, like Ashley before him, felt the tremor of fear that radiates from the rocks at the Gates of Lodore. Powell's journals see the gates as "a dark portal to a region of gloom." He and his men, though, would have more reason to fear their fate than the men who drifted in the bull boats with Ashley almost a half-century before.

Unlike the rafts today which are rowed by a single boatman, the design of Sumner's for the Powell boats kept three men employed. Two sat with their backs to the waves, each plying on a set of oars to keep the boat moving faster than the current, which helped the third man, standing at the stern, steer the craft with a long sweep oar. Powell, who had lost an arm in the Civil War's Battle of Shiloh, devised a complicated series of signals to be given to the steersmen as they approached whitewater. The signals, like the boats, proved less than effective. Sumner, in his account of the journey, had this to say about the signals: "As Major Powell was the only free-handed man in the outfit, he was supposed always to attend to that

part of the business, but I fear he got too badly rattled to attend to it properly on several occasions, notably so at Disaster Falls in Lodore Canyon."

At Disaster Falls, Powell had landed first above a drop that was difficult to see from the river, and had given the responsibility of the signals to one of the men in the boat as he walked downstream inspecting the difficulties. The Maid of the Canyon and Kitty Clyde's Sister both landed safely next to Powell's Emma Dean above the falls, but the Howland boys and Frank Goodman in the No Name missed the pull-out and were swept into the teeth of the rapids. The boat survived the first small drop and then was turned sideways, gnashing into the rocks. It quickly broke into driftwood, tossing the three men overboard. One man was able to make the shore while the other two were nearly swallowed up by the murky water, their heads appearing and then vanishing among the waves only to reappear again farther downstream. By a stroke of luck that only river runners can appreciate, there was a small gravel bar in the midst of the frenzy and the two men clung to it for their lives. Some plying at the oars "right skillfully" by Sumner swept them safely from the island and back to shore where Powell, after greeting them "as if they had been on a voyage around the world and wrecked on a distant coast," lit into them with a fury for missing the signals. They said no signals had been given, but the fire had been sparked and the mistake which cost them the No Name burned into the Major's gut like a glowing ember and began a flurry of arguments which would eventually take a heavy toll on the expedition's morale, something more valuable than the flour and bacon lost with the No Name.

Pike led his men over the Sangre de Cristo Mountains at Medano Pass and came upon the wind-whipped waves of the Great Sand Dunes, today protected in Great Sand Dunes National Monument. WILLARD CLAY.

The incident and subsequent arguments hung over the trip like a dark cloud. One of the men, Frank Goodman, left the expedition at the mouth of the Uinta River only a few days after the Disaster Falls wreck. Much farther downstream, in the heart of the Grand Canyon, the anger again came to a head. Standing above a rapids Powell himself described as "worse than any we have yet met in all its [the river's] course," the thought of Disaster Falls and their narrow escape there echoed in the crew's minds and three of the men, including the last two remaining crew members of the long-lost No Name, decided they had had enough.

Despite urges by Powell and particularly by Sumner, who had devised a skillful plan for getting safely past the new rapids, the three men gathered their gear, said some tearful goodbyes, picked up a plate of biscuits left on a rock for them by the cook Billy, and began their climb to the canyon's North Rim.

Powell, Sumner, and the remainder of the crew ran the rapids safely and pulled up in an eddy to fire their guns, hoping the sound would bring the three back to the river. It did not, and O.G. Howland, Seneca Howland, and William Dunn were never again seen alive. Ironically, just two days later the boats drifted out of the canyon, out of danger, and the world opened its arms to the crew that had run the Grand Canyon. Even in their celebration (recorded in Powell's journal), they couldn't help but wonder at the fate of the three who left: "Are they wandering in those depths, unable to find a way out? Are they searching over the desert lands above for water? Or are they nearing the settlements?" A small, bronze plaque now bolted to the canyon's north wall at Separation Canyon tells the tale: "Killed By The Indians."

Had the expedition had better and more reliable information about the river and its rapids; had the boats been designed for whitewater; had the signals worked so that the No Name made the pull-out above Disaster Falls; had the trip gone differently in Lodore Canyon, perhaps history would have recorded a different ending to the Powell journey and Separation Rapids would have a different, less fateful name. As it stands, the incident was the only true chapter of disaster in the otherwise successful and heroic story of exploration written by Major John Wesely Powell and his men.

Above: The Durango-to-Silverton narrow gauge railroad hauled ore when it first came up the Animas River in 1881. A hundred years later it carries gold of another kind—tourists—on the only regularly scheduled run of a narrow gauge railroad remaining in the United States. GAIL DOHRMANN.

Right: Red Mountain and the Idarado Mine stand sentry over a small stream which cascades out of the San Juan Mountains. The lure of ores and precious metals gave major impetus to the settlement of Colorado, first attracting large numbers of miners who in turn attracted merchants, builders, industries, and railroads to cater to the growing population. LARRY ULRICH.

Last of a breed

It is said that he didn't talk much—that trait can come from so much time spent with only canyon walls for company. He was more at home with a roof of silver stars over his head and a spot of sand along the river for a camp than in any four-walled structure.

Like the mountain men of seventy-five years before, he made do with what the wilderness yielded—beaver pelts, flecks of gold in his pan, and what game he could bring down. River historians like Otis Marston called him "one of the greatest outdoorsmen and boatmen of them all."

Hearing praise like that, Nathaniel Galloway would have humbly averted his eyes, maybe spotting rings on the water where the trout were rising and thinking about fixing camp before dark. Galloway was the last of a breed, a mountain man who ran rivers.

In 1891, Galloway heard stories around the campfire of some others who were having a bit of luck with boats modified from the Powell design, and it gave him an idea. By 1895, he traded in his pack animals for a boat and began a life on the water. In that year, he made his first long trip, solo from Green River, Wyoming to Lee's Ferry just above the Grand Canyon, trapping and prospecting as he went. In 1896, he made several trips through the canyons of the upper Green and, on one of those trips, came upon a pair of down-and-out prospectors below Ashley Falls. Galloway talked one of the men into joining him and in early 1897 the pair showed up in Needles, California, having quietly run the Colorado River through the Grand Canyon.

In 1901, the White River flowing across northwestern Colorado and the Piceance Basin caught his eye and he loaded up a rifle, his traps, his prospecting gear, and his son for a look. The trip, of which there is no further record, is believed to be the first run of the White River. Four years later, he and his son Parley, who would later join Clyde Eddy on a trip through the Grand Canyon with "3 stout boats, eleven men, a bear and a dog," put in at Lily Park and rowed on one of the earliest Yampa River trips.

Galloway's reputation grew, despite his quiet way and his lack of record-keeping or fireside yarn-spinning. He was commissioned by industry mogul Julius Stone for what had to be one of the earliest "commercial" river trips on record. Stone and Galloway followed the Green and Colorado Rivers through the Grand Canyon, making Galloway the first man in river-running history to successfully row the Grand Canyon twice. In tribute, near mile 132 of the Grand Canyon, a pair of deep and beautiful canyons meet the river—Stone Canyon and Galloway Canyon.

Still, it was not *what* rivers Nathaniel Galloway ran which earned him such a lasting place in river-running history, but *how* he ran them. He turned the practical eye of the mountain man on river running. Galloway's boats were everything that Powell's were not—light, fast on the water, blunted at the ends to shed waves and remain stable, turned up or "rockered" for maneuverability even in high waves. In short, while the early notion in boat-building was to make them strong enough to bounce off rocks, the boats rowed by Galloway were designed to be nimble enough to avoid the rocks. These were boats built with the river rather than the rocks in mind.

And Galloway did one more thing. It was a simple but radical move that revolutionized the river world. He turned around. Instead of rowing with his back to the obstacles downstream like Powell's men were forced to do, Galloway changed the design of the boat to allow him to face into the rapids and see what was coming. The trade-off for visibility was power, but the compensating factor was finesse at the oars. No longer would boatmen fight the river; the new way called for the grace of good boatmanship and a keen sense of the ways of the water to get safely through the rapids.

With this technique, Galloway ran many of the rapids Powell and others were forced to portage. In just an hour or so on the water he could cover what took Powell a day's labor. River running became fun. After running Red Canyon, including the rapids where Ashley had painted his name on the rock, he commented that the trip was "smooth sailing to a man accustomed to navigate rough water." The simple act of turning around and rethinking the river's ways took river running out of the stone age and into the modern age. Rowing a river became less a battle and more a ballet. Even today, boatmen facing downstream as they enter a rapids are said to be using the "Galloway Position."

In 1912, Galloway retired to Vernal, Utah, tying his boat to the end of an era and looking on as a new era of river running was already beginning.

Because it's there

Powell came in the name of Science; Escalante came in the name of God. Ashley and Galloway were searching for riches; Pike for a river that wasn't there. By the time the rivers of Colorado flowed into the twentieth century, there was another reason for exploration—adventure. A river, like George Leigh-Mallory's Mount Everest, could be challenged simply "because it is there."

The Gunnison River in the Black Canyon was "there" for Ellsworth Kolb, and from 1916 to 1918 he made four attempts at running what he called the "most perilous water trip in the West." He went through five boats, one set of "air tight swim suits," and "fighting death in the angry whirlpools" with four different teams of men, including on the second attempt Bert Loper, known as the "Grand Old Man of the Grand Canyon." The canyon turned them back each time, the third trip smashing Kolb's kneecap and stranding him on a ledge unable to climb out of the canyon for fourteen days and giving him a long time to consider the power inherent in such a place. Finally, beginning for a last time downstream from the section which had stopped him on the other trips, he rowed out of the tail end of the canyon which had so obsessed him, coming to the confluence of the North Fork where the river goes quiet again. The adventure was over for Kolb. But for the boaters who paddled the decades from Kolb's time to the present, the Black Canyon of the Gunnison has continued to be one of the great challenges of Colorado whitewater. Despite the three dams which have harnessed the Gunnison, the Canyon remains "there" for the adventurers of today.

The most publicized adventure of its time was the "first" run of the Yampa River in 1928, sponsored by the *Denver Post*. At best the trip was the second, and more likely the third, run of the Yampa. But for thousands of readers who followed the exploits of that four-man team, the trip undoubtedly was a first glimpse of river running. Headlines like "Two Barely Escape Death" and "Death Faced Many Times" portrayed river runners on the Yampa (thirty-seven years before the flood which created Warm Springs Rapids, by far the worst on the river) as daring, death-defying thrill-seekers.

Thirty years later, more experienced river adventurers were still trying to live down what they considered a sensationalized reputation. Grand Canyon outfitter and guide Martin Litton, for example, wrote a series of articles published in the *Los Angeles Times* in 1953 which showed "women and even children" floating the Yampa River. Litton's photographs depicted a different and more characteristic side of the Yampa River than the dramatic front-page photo of the wreck of the Leakin' Lena, which accompanied *Denver Post* headlines in 1928.

There have been, of course, true "first" runs. One of the last such runs of an entire river in Colorado was the 1948 conquest of the Dolores River by Otis Marston and Preston Walker, their two wives, and a black dog named Ditty. Their trip was both a first and also a kind of last run. For as incidents like Powell's problems at Disaster Falls with the No Name proved, and as the front-page photo of the *Denver Post* depicting the fate of the Leakin' Lena graphically showed, the wooden row boat had come to the edge of its horizons when it met whitewater.

As more and more rivers were run for their entire lengths, the frontiers of adventure changed to focus on specific canyons and rapids which had not yet been boated and for those, the old ways just wouldn't do. The wooden San Juan boats used by the 1948 Dolores run would be some of the last of their kind to make history.

Already a decade before the Dolores run, things had begun to change. In 1938, a new day was dawning—or more accurately, being inflated. Amos Berg and Buzz Holmstrom were the first to row an inflatable army surplus raft on a river in the state, pushing off into the current of the Green River above Lodore Canyon. The craft was sixteen-feet long, weighed eighty-three pounds, and must have borne a strange resemblance to a sun-dried buffalo bladder to all those whose vision of "boat" did not expand beyond the hard-cased shell of a Cataract boat. But it worked well, taking the rapids of the Green and Colorado like a bubble on a breeze, and became the first of its kind through the Grand Canyon and the forerunner of the inflatable flotilla which can be seen on any summer day in Colorado's rivers.

As much as the early history of river running in Colorado had been written in such materials as willow sticks and buffalo hide, canvas and oak planks, the modern frontier is the grounds of such exotic concoctions as ABS plastic, fiberglass, Kevlar, polyethylene, and a roll call of boats with names like Spirit, Willowa, Dancer, Blue Hole, Argonaut, Spider, Hydra, C-2, and Squirt. It was a long way from bull boats to Squirts, and at least a part of the path led through Colorado.

Running the unrunnable

One of the forces which hones the skills of both paddler and boat designer is heat, the heat of competition. The same year Marston and crew were rowing their wooden San Juan boats down the Dolores on one of the last trips of its kind, the future was beginning on a stretch of the Arkansas River just south and east of Salida. The first annual

Green River paddlers are dwarfed by the Mitten Park fault, where layers of rock—originally laid down in neat horizontal layers—were tilted skyward. Such faults—where the restless earth has shuffled the deck of rock layers—are common along Colorado's rivers.
SHERPA—SAN.

Always looking around the next bend / 43

Left: Every bend in a river can hold another surprise—like this one. JOHN TELFORD.

Top right: Another modern river explorer, a kayaker, dives into Zoom Flume Rapid in Brown's Canyon on the Arkansas River. DOUG LEE/TOM STACK AND ASSOCIATES.

Above: These rafters through Brown's Canyon on the Arkansas River learn how Seidel's Suckhole got its name. DOUG LEE/TOM STACK AND ASSOCIATES.

Left: More than 100,000 people each year float the Colorado River within the state of Colorado, drawn by the challenges of such whitewater stretches as the South Canyon rapids. DOUG LEE/TOM STACK AND ASSOCIATES.

Salida-Cotapaxi Whitewater Marathon was held in 1948. It is the oldest whitewater kayak competition in the nation, drawing world-class paddlers from around the globe. Kayak racing has become a popular sport, especially in Europe and the eastern United States. Colorado has hosted such competitions as the U.S. Wildwater National Championships in 1980 on the Arkansas River, the annual Poudre River Wildwater Races, the Woody Creek Wildwater Races on the Roaring Fork, the Crystal River Slalom, and other annual and one-time events which introduced world-class paddlers to Colorado whitewater. Some of them stayed.

Walter Kirschbaum was the 1958 World Slalom champion and had many other racing honors to his credit. In the late 1950s he gave up challenging the clock for the frontiers of "unrunnable" rapids. In 1957 he paddled solo through Pine Creek Rapids on the upper Arkansas, a rapids which has become a kind of test for expert kayakers, a test in which failure has meant death to some, including one pair of rafters who attempted it without the proper safety precautions in 1975.

Kirschbaum was in a class by himself at the time and, since no one could match his skill, many of his trips were made solo, adding to the danger. One such trip was his 1962 run of Gore Canyon on the upper Colorado. Gore Canyon, along with the Black Canyon of the Gunnison and a poker-hand full of individual rapids, is at the cutting edge of whitewater in the state. The four-mile canyon drops four-hundred feet, the brunt of that in the heart of the canyon where the river tumbles two hundred feet in one mile, squeezed like a fist between sharp, black canyon walls.

The falls of Gore Canyon ripped and collapsed Kirschbaum's boat, but by stuffing the ends full of tree branches, he was able to continue his attempt. Like so many of the first runs in the state, only a scant record survives and fireside stories have been known to be more than a little unreliable.

Some say Kirschbaum was forced to portage at

Evolution of the river boat. Top left: The bull boat, easily and quickly made of branches and hide, waterproofed with pine tar pitch or tallow and ashes. Top right: The boat used by the Powell Expedition of 1869, with two oarsmen facing backwards and a sternsman steering with a long sweep oar. Middle left: The boat developed by Nathaniel T. Galloway, featuring rockered ends, square bow and stern, and single oarsman faced forward. Above: The first inflatable raft to run the Colorado River through the Grand Canyon, a trip which also included a run of the Green River through Lodore Canyon, in 1938. Right: The latest in river-running technology—a double-tubed inflatable raft such as the "Argonaut," the first to successfully run Barrel Springs.

There are still rivers to explore and discoveries to be made along the rivers of Colorado. Here, in April, 1985, Eric Bader drops into the first wave of Barrel Springs Rapids, beginning the first-ever run of this rapids on the Colorado River in Glenwood Canyon. JEFF RENNICKE.

least two drops in Gore Canyon and that Roger Paris, a two-time World Slalom and three-time U.S. National kayaking champion was the first to run all the teeth of Gore Canyon. Still others give the honor to Dr. Walt Blackadar, an anomaly in the word of modern kayaking who came not from a background of competition but out of nowhere, discovering the beauty and art of paddling relatively late in life. Like a star that falls in the night sky, his career was short—ending prematurely in a fatal logjam which caught and held his kayak. But his mark, like the marks of others—Kirschbaum, Paris, Rob Wise, Fletcher Anderson, Ben Harding, Eric Bader, Ron Mason, and other Colorado paddlers for whom "because it's there" is reason enough—had been left on river running, affirming it as an artform and life itself as an adventure.

Anatomy of a first run

They have been called Upper and Lower Death, a pair of Class VI rapids—rated most dangerous on an international whitewater scale—that hit the Colorado River in Glenwood Canyon like a one-two punch. In 1961, five boaters from Aspen pushed off above the upper drop in a thirty-three-foot pontoon boat. The huge rubber raft came to the edge, teetered there for a sickening moment, and dropped into the fury. In seconds the raft was flipped end-over-end and shredded on the sharp rocks. Four of the five men died. For almost a quarter-century, no one tried again.

Then, on April 30, 1985, a group of professional boaters led by twenty-four-year-old Eric Bader of Boulder, Colorado, a member of the 1981 U.S. National Team, stood at the brink of the upper drop. Their credentials were impressive. All were trained professional boatmen for the Boulder Outdoor Center, licensed in Swiftwater Rescue by Rescue-3 and the U.S. Lifesaving Association, certified in First Aid and CPR by the American Red Cross, and veterans of most of the West's bigwater runs. Their equipment was state-of-the-art—the double-tubed "Argonaut" raft which was assembled beside them was the prototype of designer Kris Walker, who had also come to the rapids to watch his boat in action. Television cameras and newspaper reporters hovered at Eric's shoulder. This was not to be the kind of amateur stunt which had cost the lives of the Aspen boaters. This was a group of experts, looking to go safely beyond the present boundaries of river running. They stood on the edge of the upper drop, staring into the eyes of the "unrunnable," considering.

Then, in an act almost as courageous as running the rapids, Bader and his headboatman John Mottl turned away. The line had been drawn. They would not challenge that first drop. It would remain just a bit beyond the bend. But downstream the second drop was waiting.

Barrel Springs is the common name for the second of the one-two punch. At low water levels, it had been run by kayaks, first in 1976 by Billy Ward and Brent Brown. But never before had it been run in a raft or even kayaked at anywhere near the 3,000-cubic-feet-per-second of river which screamed through the river on this day. Bader and his group faced another, if less radical, re-drawing of the lines—a "first" run by raft at a higher flow level than had ever been run even by kayak.

Barrel Springs is different than the famous upper drop, as different as life and death. While the first is a quick and powerful surge, Barrel Springs is a long series of shorter falls and rocky channels, any one of which can flip a raft or kayaker. Standing beside each of the dangerous spots in turn, Bader and his crew discussed their route. The river roared past so quickly that its motions fused together like an endless roll of thunder. With rescuers in place along the left bank with safety ropes known as "throw bags" that can reach a person thrown out of a raft and into the river, with television cameras ready, and with Fletcher Anderson, another river pioneer, watching from the banks, Bader was ready.

There is a tension that breaks like a tattered rope once the boat is spotted at the lip of the first drop, committed to its course. Running rapids is a gamble each time, a new toss of the dice. Unlike mountain climbing, where alternative routes can be tried on the same climb by simply backing down, a rapids is a one-shot deal, the muzzle-loader of adventure sports, an all-or-nothing proposition with no reloading.

Bader had the sky-blue Argonaut "in the line" and dropped easily over the first waves. The fury was in the crux of the S-turn and there a wave spun the boat like a top, forcing Bader to row backwards towards a pair of rocks centered like a gunsight in the river's course. He split the chute perfectly and spun the boat forward where another claw-like wave grabbed and twirled the boat a second time. When Bader finally straightened out again, Barrel Springs had been run.

The challenges are different today. Instead of whole continents to explore or unknown canyons to etch on the maps, the unchartered territory lies within the waves and rocks of a single rapids. The exploration is as much of the human spirit as of the terrain. The lines which were drawn by Pike, Powell, Fremont, and Long helped to define our continent. The lines drawn by the explorers of today are helping to define the geography of human skill. There are still rapids in Colorado, like the upper drop in Glenwood Canyon, which remain unrun—undiscovered. Perhaps there needs to be boundaries remaining, boundaries which will someday be the challenge of a new generation of river explorers seeking to move ahead, if only for a look around the bend.

For the world to see

In 1870, a young photographer from New York rode into Colorado, set his camera atop the long legs of a tripod, and clicked the shutter on the first of his more than eighty thousand pictures of the Rocky Mountains. The world's eyes were opened to the West.

The photography of William Henry Jackson was not the first of its kind in Colorado. But in shows and galleries across the world, it was the work of William Henry Jackson that revealed to the world such wonders as Mount of the Holy Cross, Mesa Verde ruins, Yellowstone Falls, a geyser named Old Faithful, and many of the rivers in Colorado.

As official photographer of the Hayden Surveys of 1870 to 1879, Jackson combed the Rocky Mountains with a pair of mules carrying his three hundred pounds of camera equipment. His lens found the Royal Gorge of the Arkansas, the Black Canyon of the Gunnison, the Animas, the Colorado in Glenwood and Gore canyons, the Blue, the Rio Grande, the Mancos, and the Crystal River Canyon—where one mule lost its footing and smashed dozens of glass plates, forcing Jackson to retrace his steps and re-shoot the photographs. He would picture the South Platte already laboring under a network of irrigation flumes and would capture many of the first pictures of the railroad's conquering of the deep river canyons.

Photography in those days was no easy matter. Jackson worked with the wet plate process. After a scene was located, the mule train was brought as close as possible and the equipment hand-carried the rest of the distance. There, the bulky camera and a portable dark room were set up, the heavy glass plates prepared, and the exposure made. Then the work began. Because of the method, the print had to be developed on the spot within thirty minutes or be lost. Often Jackson would have to scramble out of a canyon and rush to his darkroom tent to soak the glass plate in barrels of chemicals also carried on the mules. Each print had to be developed, fixed, and varnished on site, a process which took thirty minutes per exposure. Using this method, the energetic Jackson shot up to thirty-two prints per day, rushing back each time to his darkroom tent. But when he emerged a half hour later, another scene was captured on glass for the world to see.

Above: Fishing was already a popular sport on Colorado rivers when Jackson found these well-dressed anglers on the Rio Grande near Wagon Wheel Gap.
Above right: Jackson pioneered photographic techniques, slow exposure here transforming the San Juan into fog.

Right: The Animas River Canyon, threaded by tracks of the Denver-Rio Grande Railroad, was one of Jackson's favorite photographic topics.
Far right: Jackson worked in later years as company photographer for the railroad. From a specially outfitted railroad car, he photographed this locomotive bellowing smoke deep in the Royal Gorge of the Arkansas River.

PHOTOS COURTESY OF COLORADO HISTORICAL SOCIETY

Chapter three:

The rivers, a tale of twenty-five

Rivers are an elusive bunch. Even Colorado can't agree on how many rivers Colorado has or what to call them. To some local old-timers and even on some maps, the river above the town of Yampa is known as the Bear River. To others, it is the Yampa River from its headwaters to its mouth. On some maps, the St. Vrain is a river and on some maps it's a creek. There are two Cimarrons and two Roaring Forks. There are two Fall Rivers, one of which muddies the picture even further by flowing not into another and larger river as most most rivers do, but into Clear Creek, raising questions about when a creek stops being a creek and becomes a river.

With such disagreement, a river count becomes no simple matter. There are four great river systems which claim Colorado as a source—the Platte, Arkansas, Rio Grande, and Colorado. Add the direct tributaries which meet these four rivers somewhere in the state and the river count rises quickly to over a dozen. Add feeder rivers of these tributaries and the number tops twenty-five. Throw in rivers beginning on Colorado soil but flowing into one of those four great rivers somewhere outside the state—the Dolores, which meets the Colorado just across the Utah line, or the Encampment, which flows into the North Platte a stone's throw into Wyoming—and add *their* tributary and feeder streams, and the total nears a hundred. Finally, consider the dozens of named forks of these rivers, the "south" or "middle" or "north" forks, and the Colorado river count comes to over a hundred.

These are the tales of twenty-five of the state's rivers and their major tributaries, a collage of the history—human and natural—which helps to paint the way we perceive rivers in our landscape. Not all the rivers are here. Not all the stories are here. They couldn't be—rivers are an elusive bunch.

Animas River

Source: Houghton Mountain in the San Juan Range.
Mouth: San Juan River near Farmington, New Mexico.
Length: 110 miles (85 miles in Colorado).
Counties: San Juan, La Plata, Hinsdale.
Cities: Silverton, Durango.
Access: U.S. 550, Colorado 789.

It sounds, on a still night, like a distant whisper or a rattle of chains when the wind is right—the rustling waters of the two creeks which come together and mark the beginning of the Animas River at a ghost town called Animas Forks. It is fitting, even poetic that the river called "spirit" or "soul" by the Spanish should rise up at a ghost town. Yet the silence of Animas Forks is both a symbol of the river's past and a key to its recent popularity.

Silver spurred the beginning of settlement in the Animas River Valley, or at least the talk of silver. Charles Baker, a self-confessed adventurer and talented promoter, recruited settlers from mines at California Gulch on the Arkansas with stories the Indians had told him of great shiny-colored rocks near the headwaters of another river: the Animas. Soon the city of Animas Forks was established at what was called Baker's Park.

The stories, to Baker's credit, were not far off and the Animas River Valley did enjoy some success, enough to lure a spur track from the Denver & Rio Grande Railroad. By 1882, the narrow-gauge locomotive had conquered the steep canyons of the Animas to begin hauling ore to mills at Durango. All told, more than $42 million in gold

Sunset at Sheep Lakes reflects the cliffs of Fall River Canyon, part of the Big Thompson drainage located in the Mummy Range of Rocky Mountain National Park.
KENT AND DONNA DANNEN.

Glacier Creek, in Glacier Basin of Rocky Mountain National Park, highlights the September morning splashes of color which embellish even the smallest, least-reknowned of Colorado's streams and pools
JEFF GNASS.

and silver were gouged out of the rock along the Animas from 1874 to 1923, enough to keep the railroad operating and to draw permanent settlers to the valley. Despite the richness of the strike, however, it has always been overshadowed in Colorado mining history by the fantastically-rich Cripple Creek and Blackhawk Gulch discoveries.

Then the silver market softened, the competition from the newly-plotted town of Silverton just downstream stiffened, and Animas Forks faded. Interest in the town as an historic site has kept it on some maps, but today only memories, winds, and curious tourists walk the townsite.

Silverton and Durango suffered, too, from the silver market crash just before the turn of the century, but have survived to take advantage of another of the riches of the Animas River Valley: its beauty. Nestled snugly in the mountain valley, surrounded by the jagged peaks of the Needles Range and dropping over two thousand vertical feet from its headwaters to the town of Durango, the Animas has cut one of the longest and deepest canyons in the Rocky Mountains. It is certainly one of the most beautiful.

It is beauty which can be seen at a glance. Driving U.S. 550, the famous Million Dollar Highway, hundreds of thousands of tourists each year get their first view of the Animas River—at a glance. The highway twists and turns its way through the rugged terrain to Electra Lake, a popular picnic area and campground. It then continues north to the town of Silverton. For most of its route the highway keeps to the less-steep grades on the rim of the canyon, far above the river, offering only an occasional and quick glimpse of the waters below.

There is another way to see the Animas, closer up. The railroad which once chugged along the river with its cargo of precious ore has been overhauled to carry tourists. Its route hugs the river, coming at times within a few feet of the water, through all but the most impassable sections, giving passengers not only a closer look at the river but also a thrilling and nostalgic canyon-ledge ride on the only regularly-scheduled, narrow-gauge railroad operating in the country.

Yet, even the train does not expose all the secrets. The Animas is a wilderness river, tucked in a kind of preservation-by-isolation cradle of its own deep canyons. For a few miles below Elks Park and again near Electra Lake, the river forms the western boundary of the 467,000-acre Weminuche Wilderness, the largest of the state's designated wilderness areas. Much of its upper route winds through lands governed by the San Juan National Forest, and its lower reaches are bordered by the lands of the Southern Ute Indian Reservation.

These wild borders have kept the Animas a free-flowing river. The Tacoma Powerplant near Canyon Creek, providing electricity for Durango, uses water flowing from Electra Lake in a series of tunnels. The Animas River itself runs free.

All of that could change. In 1968, the second-largest Colorado water project authorized by Congress between 1964 and 1972 received tentative approval. The Animas-La Plata project would divert 200,000 acre-feet annually for use in rapidly growing areas of southwestern Colorado. Support for the project has been mixed and, in recent years, Congress has had its hands full with construction of the McPhee Project on the Dolores River and has appropriated no funds for the Animas-La Plata proposal. Construction, therefore, seems to be a long way off, but the battle rages on.

Its scenery, quality fishing, wild shores, hiking trails, and challenging whitewater have made it one of the most popular rivers in the state. That popularity has brought change to the Animas River. It is no longer the path to riches in silver and gold that it once was, and the town of Animas Forks is now all but forgotten. But some things don't change, and on a quiet night when the wind is right, there are still sounds like whispers from where the Animas rises.

The search for gold and silver first brought settlers to the Animas River Valley. Today, remnants like this mining footbridge remain south of the ghost town of Animas Forks. JAMES FRANK.

Arikaree River

Source: Confluence of North and South Forks of the Arikaree River.
Mouth: North Fork of Republican River at Haigler, Nebraska.
Length: 129 miles (80 miles in Colorado).
Counties: Washington, Yuma.
Cities: Cope, Beecher Island.
Access: U.S. 385, U.S. 36, Colorado 59.

The face of some Colorado rivers would be unrecognizable to those who traveled on their shores just a century ago. The Arikaree River is not one of them. Time, like the solid boulders of the river's shores, has stood quite still along the Arikaree. A flash of lightning from the praire sky lashing out in front of a thunderstorm can still make the river rear up and run strong. The low sounds of dry prairie winds can still lull it to sleep during long droughts. There are no international resorts along its shore, no postcard pictures of its scenic wonders. It is just a prairie river that, like an old face, never seems to change.

Far downstream, just before the Arikaree leaves Colorado for Kansas and Nebraska, the waters split around a large island. The maps show it as Beecher Island. Some of the most heroic soldiers in Colorado history fought here.

Near Beecher Island, a ragged group of fifty "scouts," who were really just a group of untrained civilian volunteers assembled to squelch a recent string of Indian raids, found themselves suddenly under attack by "thousands" of Cheyennes led by the famous war chief Roman Nose. The date was September 17, 1868.

Colonel G. A. Forsythe, commander of the rag-tag troop, ordered the men to take up a strategic position on the island in the Arikaree River and dig in for a fight. A line of barricades was erected, using driftwood snagged on the island and the carcasses of horses already killed in the battle. Behind the makeshift stronghold, the men were ordered to fire sparingly in an attempt to save scant ammunition, and they alternated firing and loading to present continuous firepower.

Entrenched on the island, the fifty raw recruits held off wave after wave of attacks, and the battle dragged on day and night. Ammunition grew short. The "scouts" found themselves battling their own fatigue as desperately as they battled the Indians.

Finally, after days, a messenger was able to break through the enemy line and get word to the all-Negro Tenth Cavalry stationed at Fort Wallace in western Kansas. Help arrived on September 27, ten days after the first shots were fired, and still the dogged troops were holding off the Indians.

When the smoke finally cleared, the casualty list showed seventy-five Cheyennes killed and five dead U.S. soldiers, including the second-in-

A name sampler

Colorado—After bearing more than nine names, the name which in Spanish translated to "the color of red" was given the silt-laden Colorado River.

Uncompahgre—One of the few Colorado rivers to retain its original Indian name, in this case a Ute word for "hotwater springs."

Conejos—A Spanish name for "rabbit."

La Plata—A Spanish name for "silver."

Gunnison—Named in honor of one of its earliest explorers, Captain John W. Gunnison who led an expedition to its banks in 1853.

Yampa—Although called the "Bear River" in its upper reaches even by local residents today, the river is officially named after a plant ground up and used by the Utes for food and medicine.

Fraser—Named for an officer on the expedition of Captain Berthoud in 1860.

Eagle—A translation of an Indian word said to mean that the river has as many tributaries as feathers in an eagle's tail.

San Juan—Once called the "Powuska" or the Mad River by the Navajo, the river was renamed in honor of St. John the Baptist by the Dominquez/Escalante Expedition of 1776.

Fryingpan—The name derives from a story in which a fryingpan is used to mark a spot along the river where a man injured in an Indian raid lies waiting for help to return.

Williams—Named after Bill Williams, a fur trapper and guide for many expeditions which explored the rivers of the Rockies.

Cache la Poudre—A French name designating this stream as the "Cache of the Powder" where gunpowder was hidden.

Platte—Both the North and the South Platte were named by a pair of Frenchmen, called the Riviere de Plat, or Flat River.

Dolores—Named by the Dominguez/Escalante Expedition as "El Rio de Nuestra Senora de las Dolores," The River of Our Lady of Sorrows, in honor of one of the men who drowned in the Rio Grande near Big Bend.

Apishapa—Another of the rivers bearing its original Indian word, named for the foul smell of its waters under the hot summer sun. Locals call it the "Fishpan."

Arkansas—An Indian word meaning "bow on the smokey water."

Cimarron—One story says it was named by two old prospectors watching a pot of hard beans "simmer on" on the fire.

command, Lieutenant Fred Beecher. Today, his name remains on Beecher Island and lives on, with those of the fifty other brave "scouts," in the history of this small prairie river.

The soldiers of that battle, if returned to the place where they struggled for ten long days against overwhelming odds, would recognize the spot instantly. Little on the Arikaree has changed. Even the river's name remains the same.

The Arikaree's name is taken from a tribe of Indians called the Airkaras, or the Rees. The original homeland of the Rees was along the Missouri River in North Dakota, but in 1823 the tribe was banished from its native lands by the U.S. Cavalry for an alleged attack on a party of fur-traders who had invaded their hunting grounds along the Missouri River. The Rees came here, to the high table lands where the same winds blow as in their homeland and where time has that same slow gait, like a late-summer river. Even today, a century and a half later, the Rees would settle in as easily as a returning flock of geese on the banks of this river that never changes.

Arkansas River

Source: East, West, and Tennessee Creeks from Sawatch and Mosquito Ranges.
Mouth: Mississippi River at Arkansas-Mississippi state line.
Length: 1,450 miles (315 miles in Colorado).
Counties: Lake, Chaffee, Fremont, Pueblo, Otero, Bent, Prowers.
Cities: Buena Vista, Salida, Canon City, Pueblo.
Access: U.S. 24, U.S. 285, U.S. 50.

The Arkansas is a mountain river. The first of its waters flow down the slopes of some of the loftiest peaks in the Rockies, including 14,433-foot Mt. Elbert, the tallest in Colorado. In its first 125 miles, the river tumbles 5,000 vertical feet, the power in the falling water helping to carve out the

Above: Not always the placid water of the flatlands, the Arkansas—which enchanted early explorers—still provides excitement to tens of thousands of floaters each summer who add millions of dollars each year to the state's economy. KAHNWEILER/JOHNSON.
Left: Mighty thunderheads tower above the Mosquito Range while the Arkansas River winds its leisurely way through prairie grass. DAVID MUENCH.

Royal Gorge, one of the West's scenic wonders with the world's highest suspension bridge dangling 1,053 feet above the surface of the river. More than 75,000 people a year come to ride the Arkansas in rafts and kayaks through rapids like Zoom FLume, Giant Steps, and One-Through-Six deep in Browns Canyon, another of the river's handiworks. In the deep, emerald pools behind the rocks in the rapids, fishermen try their luck for trout in gold medal trout waters.

The Arkansas is a plains river. It leaves Colorado at the lowest point in the state (3,350 feet) near Holly and meanders lazily across three more states before reaching the Mississippi River at a point just 100 feet above sea level. By that time, the Arkansas, which started off in rivulets no deeper than raindrops, has become the longest and second-largest tributary of the grandfather of rivers, the Mississippi, bested in flow only by the deeper Missouri to the north.

The Arkansas is a river of contrasts. Running with the deep snows high in the Rockies, it has watered immense herds of buffalo far out on the Great Plains. It begins in the sky and runs across thirteen degrees of latitude, four states, and tags along with the Mississippi on its last six hundred miles, emptying into the Gulf of Mexico. Once, a weary group of one hundred men under Pedro Villasur stumbled cold and hungry for many miles up its banks in search of game and shelter only to be ambushed by Indians at an unguarded camp a few days later on the Platte River. Once, the Arkansas served as path and guide for the great pathfinders like Pike and Fremont and for great wagon roads like the Santa Fe Trail and the Cherokee Trail that opened the West. Bull boats, strange and foul-smelling tubs of buffalo hide stretched over willow frames, once scraped its shallows and wallowed under heavy loads of beaver pelts bound for places like Bent's Fort on the banks of the Arkansas.

Today, it is a river lined with fast-lane highways and railroads. Each day, travelers cover in minutes the miles that took wagon trails a full day. Each year, hundreds of kayakers gather to compete in the famous Fib-Ark Kayak Race from Riverside Park in Salida to the Cotapaxi Bridge, covering a 26.6-mile course in just more than two hours with their fiberglass and plastic boats weighing less than a single pack of beaver pelts. The Arkansas has changed, and with it the history of the West.

As the face of the Arkansas has changed, so has the face of Colorado. The river first came into prominence as a border between American- and Spanish-held territories following the Louisiana Purchase and again during land disputes between the Republic of Texas and the United States.

When the Pike Expedition of 1806 entered Colorado on its mission of exploring the southern reaches of the Louisiana Purchase, it came by way of the Arkansas River. The land they looked out across seemed to them a vast and barren desert. In his journal, Pike writes of "tracts of many leagues on which not a speck of vegetable matter exists." Later, the Long Expedition of 1820, which came at the end of a five-year drought, gazed across the eastern plains of Colorado from Long's Peak to Pike's Peak and scrawled "The Great American Desert" across its maps.

Today, the water from the Arkansas, and waters diverted to it in a series of transmountain diversions from the Roaring Fork, Fryingpan, and Eagle rivers and Piney Creek, irrigate one of the state's most important agricultural valleys. One-third of all the farm and ranch lands in the Arkansas Valley receive irrigation water from the river and its tributaries. Wheat, sugar beets, fruit, livestock, and other vital commodities are raised here and shipped all over the state and nation. The water of the Arkansas River has made "the Great American Desert" bloom.

Yet even that may not be all that is asked of this river. The basin encompasses less than one-third of the state and is home to twenty-five percent of the state's population in one of the fastest growing regions in the West. New industry, attracted by the quality of life and location, sprawls along the Denver-Front Range Metroplex, and the discovery of the recreational opportunities offered by the river and the wild mountain ranges it drains has created new demands on the Arkansas River.

It continues to be a river of contrasts—pristine wild lands and booming valley communities. Pike and the rest would not recognize the river that led them to new lands, but one thing remains the same. The Arkansas is still a path. The way growth is handled along its banks, how new demands and conflicts on the river are solved, will make the Arkansas River as much a path to the future of the West as it was a path to the present.

Big Thompson River

Source: Hallett Glacier at Milner Pass in Rocky Mountain National Park.
Mouth: South Platte River near Greeley, Colorado.
Length: 65 miles.
Counties: Larimer, Weld.
Cities: Estes Park, Drake, Loveland.
Access: U.S. 34, Colorado 119.

It wasn't raining at the Covered Wagon Cafe in the tiny riverside town of Drake. On the wall, the clock read not quite 6:30 p.m. and, sipping free coffee refills, customers talked of the celebration of the Colorado Centennial scheduled for the next day. A few of them, if they were sitting at the tables near the windows, might have noticed the black clouds gathering over the peaks. Three hours later, the Covered Wagon Cafe was gone.

The July 31, 1976 flood of the Big Thompson River began with little warning. The warm and sunny weekend had attracted the usual summer tourists to the resort towns along the river or onto U.S. 34 and through the canyon toward Rocky

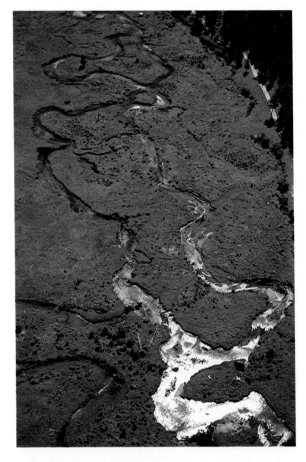

Above: The classic river profile of young, straight rivers flowing out of the mountains to meander widely across the plains does not fit the Big Thompson, which twists and turns across the high wetlands of Moraine Park in Rocky Mountain National Park, near its headwaters.
JOE ARNOLD.

Left: The Big Thompson flood of July 1976—three times worse than a hundred-year flood—ripped the heart from the river's ecosystem and killed 139 people. Five are still missing. In this photo, taken just five years after the flood, the river is still slowly recovering. JEFF GNASS.

Mountain National Park. Officials have estimated that there were 2,500 tourists in the canyon when the clouds first blocked out the sun. Over 140 of them never saw the sun again.

The rains started at 7 p.m., far upstream from Drake and the Covered Wagon Cafe. Radar used by the National Weather Bureau to track the storm indicated that 7.5 inches of rain fell between 7:30 and 8:40 p.m. The thin soils and small creeks of the Big Thompson Canyon were saturated and by 8:45 p.m. the word came—the river was flooding.

Even by that time, very little rain had fallen on towns downstream. At Drake the phones were out, but that is not unusual in the canyon. Two police officers—Sergeant W. Hugh Purdy of the Colorado State Patrol and Patrolman Michael O. Connely of the Estes Park Police Department—volunteered to drive into the canyon and warn those who could not be reached. Although the officers undoubtedly saved many lives, the flood caught them somewhere in the night.

A wall of water gathered like a fist high up in the canyon and raced downstream, growing with every swollen side creek it met. By the time it slammed into the town of Drake at 9 p.m., it was wrenching boulders as large as seven feet in diameter from its banks, hurling shells of summer homes into canyon walls, and crushing cars and motorhomes like empty tin cans. With a groan and the sound of smashing glass, the Covered Wagon Cafe was twisted off of its foundations and swept away.

A second punch hit the town when, just thirty minutes later, the North Fork of the Big Thompson River peaked at Drake. Although the river gauges were destroyed in the flood, estimates from eyewitness accounts put the flow of the two rivers at more than four times their previous high water mark. The one-two punch left thirteen dead at Drake, but the toll climbed much higher as the wall of water continued.

The flood covered the eight miles from Drake to the canyon's mouth in less than thirty minutes, peeling U.S. 34 off the canyon walls like a zipper. At a section of the canyon known as the Narrows, water rose fourteen feet in a matter of seconds, erasing the road so thoroughly that later there was not the slightest sign a major highway had stood in the canyon. At the canyon mouth, the flood began to dissipate, although damage would be extensive all along its route.

By mid-morning the following day, the sky cleared and daylight revealed the damage. One eyewitness said the devastation would "rival a combination tornado, flood and earthquake." Dozens of homes were destroyed, bridges were washed out, and 438 automobiles were totalled. All that remained of the Loveland Powerplant were solid-steel generators bolted to cement slabs.

The human toll was devastating. As rescue operations began, the death toll climbed. It reached 139 and included residents who had no time to flee their homes, tourists from seventeen states (many caught in their cars as they raced the flood downstream), and the two police officers. Five others are still listed as missing.

All is quiet now. Summer skies are mostly blue and cloudless. Unless you knew the canyon before July 1976, it is difficult to envision that anything so powerful and tragic could have occurred here. There are still signs of the disaster—boulders as big as trucks streaked with chalky scars of movement; tangles of debris jammed high into the cracks of the cliff like the nests of some great bird; the slick, new road and shiny bridges which seem all-too-new for this place. Other signs are more to the point, like warning signs advising a "climb to saftey in case of flood" and the monument erected at a gravel turnout along the highway in remembrance of the officers and others killed and bearing the words: "Lord of Hosts, Protect us yet, Lest we forget, Lest we forget."

Each summer weekend, thousands drive past the monument where the highway narrows on its way to Estes Park or Rocky Mountain National Park. The fishing is back, many of the resorts have been rebuilt on the same spots as before, and the river looks too small and rock-filled to be of much danger. Still, if a cloud passes over the sun in the late afternoon, or a jet rumbles off in the distance, the people of Drake shiver for a moment with the memory of that stormy night not so long ago when the Big Thompson River went wild. We will never forget, we will never forget.

Blue River

Source: Blue and Crystal Lakes near Quandry Peak in Tenmile Ridge.
Mouth: Colorado River near Kremmling, Colorado.
Length: 65 miles.
Counties: Summit, Grand.
Cities: Breckenridge, Dillon, Silverthorne.
Access: Colorado 9.

Towns like Dillon and Silverthorne seem caught in the middle. They are not on the Eastern Slope because they are west of the Continental Divide; yet they are not wholly a part of the West Slope because Vail Pass looms over them to the west. Being separated from the cities of the Front Range has saved them from the problems of big city life—pollution, crime, traffic—yet being only an hour's drive distant brings in many weekend fishermen, boaters, and tourists without the time to light out for points over Vail Pass and further on. It is a place in the middle.

Through the heart of this place-in-between flows a river: the Blue. As if reflecting the valley through which it flows, the Blue is also a divided river. Two large reservoirs parcel out its flow. The lower one, Green Mountain Reservoir, collects water for use by the ranchers and farmers of the

Western Slope. Further upstream and newer, Dillon Reservoir hoards water which is to be flushed beneath the Continental Divide and into the channel of the North Fork of the South Platte at the rate of 82,956 acre-feet a year. The water, used for irrigation and municipal purposes by several Front Range cities, flows through the 23.3-mile Harry D. Roberts Tunnel, completed in 1964 and at that time the longest tunnel of its kind in the world.

The Blue has long sat astride two worlds. In the early 1800s it was a favorite spot of the mountain men—it had a good supply of beaver and it lay in a neutral zone between the territories of the Osage Indians to the north and other hostile tribes further south.

Its name, which seems appropriate today, is derived from the color of its waters. Yet the Blue was not always so blue. Heavy placer mining high on the headwaters and feeder creeks of the river helped to tinge the water with the shades which earned it the name "L' Eau Bleu" from the Frenchmen who first worked the claims.

And rich claims they were. Gold samples taken from the bed of the Blue in 1860 were given the highest assay ratings for both color and purity and commanded a high price for the time at $20 an ounce.

Today, the blue waters on the lower end of the river have been given another high "gold" rating: this time, for their fish. From the Green Mountain Reservoir gates to a point 2.5 miles downstream, the Blue is a gold medal trout water. On any day of trout season, the Blue is a popular place for its

Left: Like a thread of mountain sky, the Blue River flows through high mountain valleys just an hour's drive from cities of the Front Range, attracting thousands of anglers each year with both gold medal and wild trout waters.
DAVID MUENCH.

Right: The cleanup of the South Platte in Denver has brought people back to the river, and wildlife as well. Birds like the black-crowned night heron are once again a common sight. GARY R. ZAHM/DRT.

good fishing, easy access off of Colorado Highway 9, and the wildness that still remains along its shores.

Yet, the Blue is not a wilderness river. It hasn't been for many years. Roads tag along both the mainstem of the river from its headwater to its mouth and along its major tributary, the Swan River, from its beginnings to the confluence. But the beauty remains. From many places on the river, the Gore Range and the peaks of the Eagles Nest Wilderness Area loom on the western skyline. Closer to the river, pastures watered by irrigation from the river are dotted with grazing horses from the many ranches nearby. There are several national forest campgrounds in which to spend the night, and isolated pockets hiding deep pools in which to fish for trout by day, all close enough to town for easy access. It is not an urban river and not quite a wilderness river. The Blue River, once again, is somewhere in between.

Cimarron River

Source: East slope of Raton Pass in northeastern New Mexico.
Mouth: Arkansas River near Tulsa, Oklahoma.
Length: 538 miles (15 miles in Colorado).
Counties: Baca.
Cities: None in Colorado.
Access: County and local roads only.

The Cimarron is not a boundary-drawer's kind of river. It twists and bends like a snake in the sun. If not for the boundary-drawer's penchant for straight lines and right angles, the Cimarron could have marked the state's southeastern border and made good Colorado's claim as the "Headwaters State." It is one of just two rivers flowing through Colorado which do not have their headwaters here. The other clips off the far northwestern corner of the state and was known as the *Seedskeedee* or Prairie Hen River to the Indians.

Today, it is called the Green. Despite their non-native beginnings, both the Cimarron and the Green have played important roles in the shaping of the state's history.

Today the Cimarron runs lonely and unseen for most of its miles within Colorado. Although it is a major river of the Great Plains, stretching from New Mexico to the heart of Oklahoma, its few miles in Colorado do not now play a leading role. Once, however, it had plenty of company here. The Cimarron River was a part of the great road to the West, The Sante Fe Trail. As one of the two major routes across the Rockies (the other was the Oregon Trail to the north), the Sante Fe Trail led hundreds of thousands of prospectors, mountain men, settlers, and cattlemen into the new territory of the West. Kit Carson, the man who has come to symbolize the spirit of that time in Western history, often rode the Sante Fe Trail when not leading one of Fremont's expeditions, trapping the mountain streams, or exploring unknown valleys.

The trail also was the route of claim-jumpers, rustlers, and outlaws. Like so many of the Western rivers, the Cimarron held secrets. Near its headwaters, there is a deep and maze-like canyon which hid many of the notorious outlaws and desperados of the early West. The canyon was an informal part of the same Outlaw Trail which connected places like the lower Dolores River canyons, Robber's Roost in Utah, the Powder River in Wyoming, the Browns Park hideouts, and other places where the likes of Butch Cassidy, Flat-Nose Curry, and the Hole-In-The Wall Gang laid low.

The bends of the Cimarron were the only places where the Sante Fe Trail touched Colorado soil. Along the trail came the men and women who would shape the history of Colorado, the famous and the infamous. Today, the trail is just a memory of footprints scattered by the wind, and the Cimarron flows quietly towards greater things in other states downstream.

Colorado River

Source: Middle Park in Rocky Mountain National Park.
Mouth: Gulf of California in Mexico.
Length: 1,440 miles (225 miles in Colorado).
Counties: Grand, Eagle, Garfield, Mesa.
Cities: Kremmling, Glenwood Springs, Grand Junction.
Access: U.S. 34, Interstate 70, U.S. 40, county and local roads.

They are in a small, slickrock canyon just upstream from where the Colorado River slips into Utah. The maps don't show any name for the canyon, and it has never been called anything special among the boatmen who run the river. It's just a canyon, one with a large amphitheater carved into the east cliff face and two simple stick-like figurines—pictographs—painted on one wall facing out towards the river.

The pictographs, one red and one white figure, date back to the Fremont Culture, a hunting and gathering people who inhabited this region from A.D. 500-1200 and left signs of their passing in paintings like these and petroglyphs which are chipped into the stone up and down the canyons of the Colorado River and its tributaries. In some places, there are groups of figures—elaborate depictions of desert bighorn with long, trailing horns; criss-crossed patterns of rattlesnakes; eerie, untranslatable forms; and of course these two unadorned figures silently facing the river from an unnamed canyon wall.

Man has left his mark on the Colorado River in many ways, too often much less artistically than the Fremont pictographs. Like many of the great rivers of the world, the Colorado has often been abused, polluted, and neglected, a victim of short-sighted goals, lack of long-term comprehensive planning, and the dry country through which it flows.

A comprehensive view of the Colorado River is difficult, but statistics help. It is a big river, sixth-longest of the 135 U.S. rivers. It drains 242,000 square miles of the West in seven states, 8 percent of the continental United States, and the waters flow 1,440 miles from headwaters to mouth.

The modern Colorado is a hard-working river, despite the observation by well-respected travel writer G. W. James, who wrote in 1903, "The Colorado River is unlike any other great river in the world. For present purposes it seems almost useless."

Just six years later, in 1909, those "present purposes" were being served by the Colorado River with the completion of the Shoshone Power Plant in the heat of formidable Glenwood Canyon. The plant was an engineering marvel of its time, forcing the river through a 2.8-mile-long tunnel bored through solid granite to produce electricity which was then conducted through the first 100,000-volt powerline to cross the Continental Divide, bringing power to Denver.

Since then, the Colorado has been put to use in many ways. Today, there are sixty-five dams corraling the waters of the Colorado and its tributaries, providing electricity for 5.5 million people, irrigating 7.2 percent of the nation's croplands which grow 15 percent of the produce and livestock, bringing water to the taps of 14.6 million residents, and providing an annual recreational resource for 13.6 million people, tourists and residents alike. The Colorado River is hardly "almost useless."

The trouble begins with simple arithmetic. More water is demanded of the river than the river has to give. The result is a complex network of laws, bylaws, decrees, compacts, treaties, and paperwork trying to squeeze water from stone. In dry years, the proud river ends unceremoniously in a salty mudflat miles before reaching its mouth at the Gulf of California in Mexico.

It all begins in the mountains of Colorado where

Above: The North Fork of the Colorado River weaves through the lush Kawuneeche Valley, where the river which cuts the deepest and grandest canyons on the continent is barely a jump-step wide.
GARY ROBERTSON/AMWEST.

Left: Winter beautifully recasts one of the Colorado River's major tributaries, the Eagle, in silent hues of silver and white near Avon. Colorado receives more than half its annual precipitation in snowfall. In fact, the most snowfall ever recorded in a twenty-four-hour period in North America fell on April 14, 1921 near Ward on the East Slope—75.8 inches.
DANN COFFEY/THE STOCK BROKER.

Most of the Colorado River flows outside the boundaries of the state of Colorado, and the river provides scenic recreation for residents of six other western states. Here, just across the Colorado border in Utah's Westwater Canyon, the cliffs reveal dark Precambrian rock formed nearly 2 billion years ago. SHERPA—SAN.

trickles of snowmelt from the Continental Divide become a perennial stream near Poudre Pass. Yet for many years, Colorado did not have a Colorado River. Despite the fact that the U.S. Senate record for February 4, 1861 (the day the territory of Colorado, originally called Idaho, was named) explains the reason for the name—". . . . the Colorado River arose in its mountains and there was a peculiar fitness in the name . . . and it is the handsomest name that could be given to any territory or state"—there was no "Colorado River" in that part of the territory which later became the state of Colorado.

During more than two hundred years of exploration on the river, many maps were drawn and redrawn. Surviving maps show at least nine different names for what is now the Colorado—from El Rio de Buena Guia to Firebrand River to Grand River—and until July 25, 1921, it remained the Grand River from its headwaters to its confluence with the Green River flowing from Wyoming's Wind River Mountains, where the Colorado as then named was born.

On that February day in 1861, a congressman from Colorado by the name of Edward F. Taylor received final approval on a bill changing the river's name to Colorado all the way to its headwaters in the state of Colorado. Some residents of Wyoming and Utah were less than happy with the change, since the Green River which flows through those two states is the longer of the two tributaries and so, some thought, more befitting of the honor. In any case, the bill passed and the state of Colorado got its Colorado River.

Perhaps the name was fitting in another way. The mountains and valleys of the state of Colorado make up only 10 percent of the river's basin. Yet, the state's deep snows provide almost 75 percent of the water in the river. The river and its tributaries drain one-third of the state, an area which receives 60 percent of the precipitation. Cumbres Pass on the border of the Rio Grande-Colorado River basins receives an annual average of over three hundred inches of snow. Wolf Creek Pass has recorded as much as 415 inches of snow annually. This run-off makes the Colorado Basin above the confluence with the Green the larger of the two stream basins and so deserving of the title of "headwaters" to the river that waters the West.

Near its headwaters, the river hardly seems to earn its name, a Spanish word for red. It is a clear mountain stream flowing into the glacially formed Grand Lake, a popular tourist spot once considered the source of the river.

Brilliant sunset and graceful rainbow combine for a stunning portrait of the Dolores River Canyon.
MARGARET YOUNG.

On a river renowned for its canyons, the first appears just downstream from Grand Lake. Gore Canyon is a deep, dark granitic canyon churning the Colorado River into rapids which stood unrun until the late 1970s, when they were finally conquered without portage by the legendary kayaker Walt Blackadar. The first successful run of a raft through the rapids of Gore Canyon waited until Rob Wise and crew attempted it in 1977. Ironically, one of the state's most popular boating sections on the Colorado River lies directly downstream from the treacherous rapids of Gore Canyon. More than forty thousand boaters a year come to that dusty, flat spot on the banks of the upper Colorado for the one-day boat trip on the beautiful river. In the clear waters, the fishing is good for both brown and rainbow trout, species which disappear downstream when the river turns its natural muddy color.

Most people see the Colorado beginning to take on that "too-thick-to-drink, too-thin-to-plow" consistency at Dotsero, where Interstate 70, the main route west, meets and follows the river. By that time, the Colorado has picked up the waters of the Fraser River flowing off Berthoud Pass (although today the majority of the Fraser, 58,000 acre-feet per year, is diverted through the Continental Divide to South Boulder Creek by way of the Moffat Tunnel), Williams Fork River, Blue River, and Eagle River, which joins the Colorado near Dotsero. The largest of these high tributaries, the Eagle, was said by the Indians to have tributaries of its own "as many as the feathers in an eagle's tail."

From there, the Colorado River cuts its most formidable canyon within Colorado just upstream from Glenwood Springs. Taking its name from the town at its mouth, Glenwood Canyon towers over the river for twenty miles. It is the home of a railroad, a soon-to-be-completed and controversial four-lane highway, and the Shoshone power plant.

At Glenwood Springs, the Colorado takes in the waters of the Roaring Fork and winds across broad valleys until dipping again into a canyon at DeBeque. Below, it takes on the waters of its biggest tributary in Colorado, the Gunnison River, at the largest town on the Western Slope, Grand Junction, and slides off towards the Utah state line.

Just before it reaches that line, it begins to take on the more familiar look of the Colorado, the face that most people recognize. Here, it cuts through Horsethief and Ruby canyons, the beginnings of the slickrock country. The canyon walls are the color of whiskey and as smooth as the flank of a roping horse.

This is canyon country, a big place. The two counties the Colorado flows through as it straddles the border—Mesa County in Colorado and Utah's Grand County—have a combined area six times greater than the state of Rhode Island with one-eighth the population. Here, the river is so silted that it crackles on the aluminum bottom of a canoe with a sound like frying bacon.

This is high desert country, a far shot from its beginnings in snowfields. In the days of August, hotter than a buzzard's breath, the rain can dry up into thin air just an arm's reach above the ground and vanish, a phenomenon the weathermen call "virga." Here, the Colorado River becomes the river that flows in most people's minds.

Down in that slickrock country somewhere on a riverbed, there are two simple figures painted to stare out at the river from an unnamed canyon. The Colorado has little gold in it, lying as it does off the state's mineral belt, and too few beavers on the main stream to have attracted mountain men.

It is a river which has only come into its prominence as the one commodity it does have—water—becomes more valuable than either gold or furs. The discovery of the value of the Colorado River has opened a new era on its banks and, standing down in that unnamed canyon, one cannot help but wonder just how man will leave his mark on the Colorado River this time.

Dolores River

Source: Bear Mountains in the La Plata Range, San Juan National Forest.
Mouth: Colorado River northeast of Moab, Utah.
Length: 230 miles (180 miles in Colorado).
Counties: Dolores, Montezuma, San Miguel, Montrose, Mesa.
Access: Colorado 145, Colorado 141, Colorado 90, county and local roads.

The building seems to tilt a bit from the wind, standing as it does in the open on the ledge above

Above: In 1982, the Dolores River was the fourth most popular whitewater rafting river in Colorado. With the closing of the gates of McPhee Dam in 1983, however, stretches of the wild Dolores like this one in Gateway Canyon have been harnessed. Release flows from the project may prove too low for boating.
KAHNWEILER/JOHNSON.

Left: Autumn splendor along the upper canyon of the Dolores. DAVID MUENCH.

the river. Hand-painted signs say "ice cream," "water 10 cents a gallon," and there are comic books on the magazine rack. Welcome to the Slickrock Cafe.

Once, this cafe served only local ranchers. It has stood on this windy hill and seen the coming and going of several waves of prospectors in search of one mineral or another. Today, it serves as a way-station for rafters coming downstream from the upper Dolores and planning to continue through the lower canyons between Slickrock and Bedrock. As well as any, its dogged persistence shows the grit needed to hang on through time.

If a person wanted to get so lost his shadow couldn't follow him, the Dolores River would be the place. In the lower canyons there are places where an echo can get lost coming back; and for years, in the early days, these canyons played host to a list of wranglers, rustlers, outlaws, and hermits as long as a good hanging rope. It was a less-formal member of the hangouts connected by the Outlaw Trail. Signless trails out of here led to such places as the Powder River Valley, which hid the famous Hole-in-the-Wall Gang; Brown's Hole (today called Brown's Park) on the Green, where the likes of the Wild Bunch holed up; Robber's Roost in Utah, frequented by "Flat Nose" Curry; and the Maze in what is now Canyonlands National Park. It was outlaw country, and the canyons of the Dolores River cloaked the getaway of Butch Cassidy and two of his Wild Bunch when they hit the San Miguel Valley Bank in Telluride on June 24, 1889, making off with $10,500 and earning a place in history.

Mining, a more legitimate form of getting rich, has seen several cycles of boom and bust along the Dolores River. First, silver was discovered at Rico near the headwaters in 1879. The crash of the silver market in 1893 shook Rico, and other towns which had sprung up to serve the mines. Several went under, but the discovery of coal deposits near Rico saved that town from a similar fate.

Here, where snows of the San Juan Mountains melt into the San Miguel River, Colorado's state tree flourishes. The blue spruce, a highly water-dependent species, is replaced by pinyon and juniper when the largest tributary of the Dolores enters canyon country. CARR CLIFTON.

Hanging Flume in the canyon below Bedrock. From the river level, it looks like a misplaced railroad line. From above, at the highway turn-out with a historic marker, it looks like the work of a fraternity prankster. The wooden sign at the canyon's edge tells the sad story.

In the late 1800s many gold discoveries were panning out along the Dolores, bringing with them riches for a few and big dreams for many. When several prospectors discovered what they figured to be a rich strike at the Lone Tree Placer mine downstream from San Miguel, no obstacle could disuade them from getting at their fortunes. The problem lay in the fact that the claim was four hundred feet above the Dolores River on an old river ledge, dry as crow feathers and beyond any practical means of getting water for the operation. Still, a plan was devised and construction begun on a ditch winding eight miles down the San Miguel to the confluence with the Dolores and its sheer canyon walls. There, the flume was begun with construction workers dangling four hundred feet above water level by ropes played out from the rim. Slowly the four-by-six-foot flume took shape, snaking its way down the Dolores towards the Lone Tree mine. By 1891, it was complete and sluicing operations began with high hopes. Unfortunately, after the incredible struggle and effort to build the flume, the gold at the site proved too fine to be sluiced in economic quantities. The dreams died. The engineer for the project shot himself. And the flume was left to crumble from the canyon walls like the splintered dreams of the prospectors of the Lone Tree Mine.

For a time, river running has been the new fortune on the Dolores River. Several commercial rafting companies now offer two- to five-day trips down the scenic canyons of the Dolores. In 1982, it was the fourth most popular whitewater boating river in the state with an estimated sixteen thousand user-days, an increase of nearly 40 percent from just two years earlier. Attractions like

Next, it was discovery of an odd, yellow, powder-like ore called carnotite that heralded another upswing in the fortunes of the Dolores River Valley. A sample of the ore, from which uranium is derived, was sent from near Slickrock to Paris, where it was inspected by the world-famous scientist Madam Curie who extracted radium from the ore. By the early 1900s when a number of uses were found for radium, the mines were geared up and soon were producing one-half of the world's carnotite.

Discovery of a cheaper substitute brought the mines to a close by 1923. Since then, other minerals have been uncovered—gold, iron, lead, zinc, lime, salt, uranium, vanadium, high-quality fine clay—and have swung fortune to the towns of the Dolores River for a short time.

Some mining still bolsters the economies of the local towns, but many places along the river now rely on different and more stable resources. Towns like Telluride in the upper basin have become meccas for skiers and tourists, relying on the area's natural beauty for their livelihood.

In the lower valley, the Dolores River, through irrigation, has provided ranchers and farmers the water needed to produce a wide variety of livestock and crops. Dove Creek, a small town near the river, has proclaimed itself "Pinto Bean Capitol of the World."

Pinto beans and tourism do have their share of problems and hard times, but nothing like the run of hard luck brought to mind by the sight of the

The sight of the narrow canyon at Gates of Lodore on the Green River, with its sheer canyon walls of Precambrian rock, put fear in the hearts of early expeditions. Today, it signals the start of a popular whitewater trip.
GERRY WOLFE.

Snaggletooth Rapids and the long scenic float through the canyons below Slickrock were bringing in tourists and their dollars to the economy of the region, and cafes like the one at Slickrock had found another reason for being there.

Now, however, the McPhee Project, a major dam and reservoir site near the town of Dolores, will change the flows of the river drastically. Although too early to tell what the release schedule of the dam will be or what effect it will have on the corridor of the Dolores River, one thing is certain—with the closing of the gates at the McPhee Dam, changes are coming again to the Dolores River.

But up at the Slickrock Cafe it's business as usual, taking things one day at a time.

Green River

Source: Wind River Mountains in Wyoming.
Mouth: Colorado River in Canyonlands National Park, Utah.
Length: 741 miles (39 miles in Colorado).
Counties: Moffat.
Cities: None in Colorado.
Access: Colorado 318, roads in Dinosaur National Monument.

It was some time ago, a hundred million years or so. The Rocky Mountains had not yet been uplifted. The Grand Canyon had not yet been carved. Dinosaurs ruled the living world. But some of those creatures roaming the humid, grassy swamp which then covered western Colorado were dying. Their huge bodies were being lifted by the waters of an ancient, sluggish Jurassic-period river and being carried downstream to where others were piling up on a gravel bar. There, the bodies were slowly covered with silt of the river, silt which would become the rocks of the Morrison sandstone formation.

It was some time ago, three-quarters of a century or so. To be exact, it was August, 1909. An archeologist named Earl Douglas noticed a large bone sticking out of the sandstone near the Green River along the Colorado-Utah border. It was the vertebrae of a Brontosaurus and would be the first in more than 700,000 pounds of fossilized bones from twelve species to be chipped out of the Morrison sandstone along the banks of Green River.

Those two events, so separated in time, led to the 1915 designation of an eighty-acre site surrounding the bone quarry as a National Monument and provided the heart of what became the Dinosaur National Monument when, in 1938, it was expanded to over 200,000 acres including the corridors of two of Colorado's greatest rivers—the Yampa and Green.

The Green River is the second major exception to Colorado's claim as the Headwaters State. Like the other, the Cimarron in southeastern Colorado, the Green has had a definite hand in the history of Colorado.

Far upstream from the dinosaur quarry lies a shaded, isolated park split by the modern borders of Colorado and Utah. The outlaws and mountain men knew it as Brown's Hole. Today, it is called Brown's Park.

The Hole was the most famous "rendezvous" site in the West, an annual gathering place for the otherwise solitary mountain men. Once a year, the trappers came out of the hills to trade their furs for another winter's worth of supplies. But these rendezvous were more than just a kind of early West flea-market—a place to renew old friendships and rekindle old hatreds. The mountain men brought together the most free-spirited and strong-willed men of the time, hungry for companionship, thirsty for good whiskey, longing for a good fight and a tall tale, and ready at the drop of a coonskin cap to defend to the death their honor and pride. There was trading and drinking and fighting and games with the Indians. There were shooting contests and horsemanship competitions. There was killing and stories of beaver as thick as mosquitoes and then, one by one and alone, they drifted off into the hills again until the next year.

The Brown's Hole rendezvous were not be missed, and the list of those who attended the annual events in the decade between 1830 and 1840 reads like a list of mountain legends—Jim "Ol' Gabe" Bridger, Thomas "Broken Hand" Fitzpatrick, Kit Carson, William Sublette, Bill Williams, M.T. (the initials stood for "Master Trapper," a self-appointed title), the Bible-toting Jedediah Smith, Colorado's own Jim Beckwourth, and John Colter, discoverer of the Yellowstone region. Many of these men trapped and explored the Colorado Rockies and later, when a change in European fashion took the bottom out of the beaver trade, some became guides for settlers moving into the state. Today, Brown's Park is a National Wildlife Refuge and the sounds of geese and ducks, sandhill cranes, and whistling swans echo where the air was once filled with gunshots, shouts, and tall tales. The mountain men are gone, but the names of those who gathered there live on in the names of mountains, passes, rivers, counties, cities, and canyons throughout the lands they helped open to a nation.

Downstream from the park, the Green River swings to the south and enters Dinosaur National Monument at a graceful and foreboding sweep of the canyon walls known as the Gates of Lodore. The Precambrian rock of the gates, once an omen of gloom and danger to early river explorers, is today a welcome sight marking the beginning of a popular float trip through Lodore Canyon.

No roads follow the river here and there are few signs of man. Even the jet contrails can seem like long, thin clouds if you squint hard enough. It is a place that has changed little. The stillness remains deep on the quiet water, and even the shadows of Winnies Grotto, a deep, axe-like sidecanyon sketched in the journals of the Powell Expedition more than a hundred years before, seem just as dark and cool.

Flaming Gorge Dam, upstream from the Monument, has taken some of the fight out of the river, although rapids like Disaster Falls and Hell's Half Mile can still live up to their names. It has not, however, faded the beauty or romance of Lodore and there are many sections of fast water for the boater.

One campfire tale, heard long ago on a boat trip through the canyon, says that the waves of the river in Lodore were the work of one "Queen Ann" Bassett, a local cowgirl who often made her living with "borrowed" cattle. Once, with a posse hot on her trail, Bassett drove a herd of rustled cattle over the cliffs at Zenobia Creek to avoid the law. To this day, so the story goes, the ripples on that stretch of Green River are the work of the legendary cattle forever making their way to shore.

Downstream the Green slides up against the massive mountain of slickrock called Steamboat Rock, marking the beginning of Echo Park and the confluence with its largest tributary, the Yampa. Here the river swings to the west, completing its semi-circular route through the state's northwest corner. Peregrine falcon nest high on Steamboat Rock and swoop in the wind, and the river flows quietly into Utah, the boats of rafters floating slowly like pieces of weathered driftwood or the bones of some giant dinosaur.

Gunnison River

Source: East and Taylor Rivers at Almont, Colorado.
Mouth: Colorado River at Grand Junction, Colorado.
Length: 140 miles.
Counties: Gunnison, Delta, Montrose, Mesa.
Cities: Gunnison, Sapinero, Delta, Grand Junction.
Access: U.S. 550, Colorado 92.

A confluence is a special place, a place where

Above: Even impressive cascades like the one at left are dwarfed by the stunning Black Canyon National Monument farther downstream. In fact, the 1877 Hayden Atlas, published after the Hayden Survey of that decade, contained this description: "There is nothing in America that equals the Grand Canyon of the Gunnison."
STEWART M. GREEN.

Left: A major tributary of the Gunnison River, the Lake Fork plunges deep into a crevice of the San Juan Range.
DAVID MUENCH.

two rivers flow into one, two landscapes shaking hands. One of the biggest confluences in Colorado is found just south of Grand Junction. Here, two very different worlds come together in two very different rivers—the Gunnison, begun as a pair of small rivers draining the Elk Mountains, and the Colorado, with headwaters much farther northeast in Rocky Mountain National Park. Here, their two worlds become one in a marriage of rivers.

The Gunnison itself sometimes seems like two rivers. Standing beneath the 470-foot Morrow Point Dam, the first major double-curvature concrete dam in the United States, it seems like a river of modern technology flowing to meet the demand curves for electrical power or irrigation. The waters of the Gunnison River Basin are used and re-used and then used again through the largest series of reservoirs and dams in Colorado. Mainstem dams like the triple Curecanti Project, which consists of Blue Mesa, Morrow Point, and Crystal dams, combine with facilities on tributaries like the Taylor, North Fork, and Uncompahgre to make the waters of the largest river basin in Colorado earn every foot of its descent. This Gunnison River is a product of modern technology.

The other Gunnison River is a product of more powerful, natural forces. Standing in the depths of the Black Canyon of the Gunnison staring up at the thin blue thread of the sky 2,800 feet overhead, or floating the rapids of the Gunnison Gorge where peregrine falcon slash the canyon winds, this Gunnison becomes a wilderness river flowing with its natural pull to the distant sea. The steep walls have turned back the railroads and highways. Only narrow, rugged trails, winding like rattlesnakes through blind canyons where there are deer bones scattered in the dust, give access to the river. There, deep in the canyon, is the other Gunnison River.

The first thing one notices about the Gunnison where it flows free is its water—green, the color of riverbank grass. There is an intrinsic beauty about the flow of its water over the rocks, a motion like the flames of a campfire, hypnotizing. From the river's edge, it is possible to sit for hours and gaze at the way the river bends over a rock in midstream, goes white at the rapids, deepens to a color like night sky where it slows to curl catlike against a sheer wall in an eddy. Just for the sheer artistry of its water, there is no more beautiful river in the West than Colorado's Gunnison.

To protect this river and keep both worlds represented on the Gunnison, a twenty-six-mile stretch from the Black Canyon of the Gunnison National Monument, through the lesser-known section called the Gunnison Gorge, to the confluence with the Smith Fork above the City of Delta, has been recommended by the National Park Service and the Bureau of Land Management for protection in the National Wild and Scenic Rivers System. A proposed dam at the Smith Fork could still the wilderness world of the Gunnison. The designation as a wild river could keep it intact. For now, the two worlds hang in the balance.

The most well-known of the wilderness worlds of the Gunnison is the Black Canyon of the Gunnison National Monument. Established in 1933, the Monument borders protect the thirteen-mile heart of the fifty-three-mile canyon, the deepest sections, the most awesome views, and offer the visitor eleven scenic overlooks, interpretive displays, and hiking trails into the canyon. Each year, an estimated 300,000 visitors get their first disbelieving glimpses of the Gunnison River winding like a thin, green thread two thousand feet below.

The Black canyon was known to, and mostly avoided by, the early explorers of the region. Dominguez and Escalante, in 1776, crossed the Gunnison near Delta and found a cross carved into a tree by another Spanish traveler, Juan Maria de Rivera, eleven years earlier. The pair turned north before coming to the Black Canyon, as did the namesake of the river, Captain John Gunnison, in 1853.

For most early travelers, the canyon's scenery, however dramatic, posed little interest or hope in their search for a route to other places, gold, furs, or a railroad grade.

Yet its beauty was not lost on at least one early expedition. The men of the Hayden Survey party laid eyes on some of the West's most inspiring scenery and, in the end, concluded in their Hayden Atlas of 1877 that, "There is nothing in America that equals the Grand Canyon of the Gunnison." One look into the shadow-filled canyon from above, or a day spent watching the waters of the river dance in the pools and rapids below, will show that the Hayden party was quite possibly right.

Huerfano River

Source: Creeks from the Sangre de Cristo and Wet Mountains.
Mouth: Arkansas River near Boone, Colorado.
Length: 108 miles.
Counties: Huerfano, Pueblo.
Cities: Gardner, Farisita, Badito.
Access: Interstate 25, Colorado 69.

First, it's the butte, which comes into view a long way off. Next, if you know where to look, there is the river—the Huerfano. Borrowing its name from the nearby tower of black basalt named The Orphan by an unknown Spanish traveler, the Huerfano fits its title, rising randomly from the creeks which flow, or not, given the whims of the summer rains and spring run-off.

For an overlooked river of the plains, the Huerfano has a lot of historical footprints along its shore. The most ancient belong to a tiny four-toed horse that roamed the area in the early Tertiary period (70 million to 3 million years ago). Some of the earliest known skeletal fossils of the pony-sized "dawn horse," known to science as *Hyracotherium,* have been uncovered in the Huerfano basin deposits.

More recent, though more mysterious, are the footprints of the Spanish expedition around 1800 who first saw and named the butte which gives the county, the river, and a creek their names.

The tracks began getting deeper by the time Captain John Gunnison arrived in 1853 on his railroad survey following a northern tributary of the Arkansas. His scouts came upon a well-defined trail leading from over Raton Pass from Pueblo on the Arkansas River to Fort Laramie on the Platte. Gunnison followed the trail to the Cuchara River, a main tributary of the Huerfano with headwaters in the Culebra Mountains south of its parent river. Where Gunnison passed the Cuchara, he records it as a river forty feet wide and two feet deep.

By Gunnison's time, the Huerfano Butte had become a famous landmark on the Taos Trail. Thousands of settlers set their sights on the butte and the river which flows nearby as they made their way west. Kit Carson knew of the butte, traveling the trail many times to and from Taos, New Mexico.

The footprints got more numerous when, in 1866, Charles Goodnight and his partner began large cattle drives along what became known as the Goodnight-Loving Trail. It led from their Rocky Canyon Ranch south of the Arkansas, along the Front Range, to Denver and the growing cattle towns of southern Wyoming.

The mark of travel became more permanent when a Denver & Rio Grande train made the maiden run up and over La Veta Pass of the Cuchara River Basin. As the train topped the pass, it marked more than the start of a "downhill-all-the-way" run; it made railroad history by reaching an altitude of 9,382 feet, a full thousand feet higher than any train in the world had gone at the

The lonely Huerfano River in Lost Canyon.
STEWART M. GREEN.

time the double locomotive of the D&RG did it in 1877.

Today, many of the footprints have been paved over by Colorado 10 and county roads. The nearby highways shoot the modern traveler through the Huerfano and Cuchara River valleys with hardly a second glance at the landscape. In glassed-in, air-conditioned comfort, travelers pass in minutes what took the wagon trains a full day or longer.

The dust of the trail has settled now. Gone are the long days in the saddle, and cars leave no footprints to add to the history of the valley.

Los Pinos

Source: Weminuche Pass in San Juan Mountains.
Mouth: San Juan River at Navajo Reservoir in New Mexico.
Length: 70 miles (65 miles in Colorado).
Counties: Hinsdale, La Plata.
Cities: Bayfield, Ignacio.
Access: Colorado 151, county and forest roads.

There is something distinctly different about the upper Los Pinos River, something nearly unique among Colorado rivers. There are clues in the silence, the soft sounds of water flowing, and the occasional whisper of wind through the trees.

One look at any Colorado road map will show at a glance how important a role the river valleys have played in the state's transportation system. Almost every major highway in the state follows a riverbed, and the same can be said of the routes of the railroads. It was the rivers which paved the way for travel. And few rivers in Colorado aren't followed closely by a highway or railroad tracks. The upper Los Pinos is one.

It does not escape entirely. As the river flows down past Emerald Lake and Granite Peak, nearing the Vallecito Reservoir, forest and county roads reach up to meet it and then follow the river as it winds down towards Bayfield, Ignacio, and the few towns on its course. It continues out of the San Juan National Forest and onto land belonging to the Southern Ute Indians. From then on it is like most other rivers in Colorado. But for twenty natural miles, much of it within the borders of the Weminuche Wilderness Area, the Los Pinos is as wild as any river in Colorado.

This wildness has not gone unrecognized. The Los Pinos is a favorite route of hikers and climbers into the mountains of the Weminuche Wilderness Area. The river itself was one of twelve rivers in the state studied for potential inclusion in the National Wild and Scenic Rivers System. In a rare move, the entire fifty-four-mile stretch of Los Pinos that was studied was recommended for inclusion under the "wild" category of the system, the classification with the most strict requirements. That fifty-four miles which qualifies as wild river represents more miles found eligible for that designation than on any other single river in the state. The Los Pinos is truly wild.

That is not news to anyone who has hiked along the riverbed, camped within earshot of the music of its waters, or fished for trout in its pools.

"Wild" is not a description that is put on a river like a name tag, and it cannot be created simply by writing on a map or list. It is a quality that is either there or it not. On the Los Pinos, it is there.

Like the other ten rivers found eligible for the National Wild and Scenic Rivers System, the Los Pinos must wait for official verification of its wildness. In the meantime, the Los Pinos is fortunate. With its upper reaches tucked safely in the Weminuche Wilderness Area, the river is less threatened than some. It can wait more patiently for recognition and protection, content for now with the sounds of the wind through the trees which earned it its name, first appearing in the 1877 Hayden Atlas—Los Pinos, "The Pines."

Colorado's rivers sustain more than human populations. Elk, for example, thrive in the wilder mountain sections of Colorado like the Los Pinos River watershed in the San Juan Mountains. MICHAEL S. SAMPLE.

Mancos River

Source: Hesperus Mountain in the La Plata Range.
Mouth: San Juan River near Four Corners in New Mexico.
Length: 65 miles (60 miles in Colorado).
Counties: Montezuma.
Cities: Mancos.
Access: U.S. 666, U.S. 160, Colorado 184, local roads.

Imagine the surprise of Richard Wetherill. In 1888, on the trail of some of his stray range cattle on a mesa just north and west of the Mancos River, he topped a rise and came suddenly face to face with another world. There, beneath an overhang in the cliffs below him, stood the abandoned ruins of a culture which had vanished six hundred years earlier. Wetherill had slipped through a crack in time, and as his partner rode up to join him, the two looked down into what has become the focal point for visitors today at Mesa Verde National Park in southwestern Colorado.

Ruins on the Mancos River were not wholly unknown. William Henry Jackson, the famous photographer of the 1870 Hayden Survey Party whose lenses first showed the world such scenic wonders as Colorado's own Mount of the Holy Cross, visited the Mancos River in 1874, photographing the river and nearby minor ruins. Although his pictures grabbed the interest of many, it was left for a chance encounter between a cowboy and an ancient city to open up for the world the heart of a culture which had come and gone before a sailor named Christopher Columbus was born.

The life blood of the Pueblo culture, which first appeared in the region around the time of Christ, was the water of the Mancos River watershed. Many creeks flow off the southeastern-tilting mesa to drain into the Mancos below, creating a fertile valley oasis. As the culture developed into its "Golden Age," ingenious irrigation systems were constructed which caught the run-off headed for the Mancos, stored it in large reservoirs for a time, and transported it 3.5 miles through a series of ditches where it brought life to the staple crops—corn and beans. The now-dry ruins of Mummy Lake in the national park are an example of the early irrigation systems.

That was the best of times for the Pueblo culture. Many of the more expansive ruins and elaborate artifacts date back to that period between A.D. 1100 and 1276, a time of plenty in the Mancos River Valley. Times would not always be so good.

It must have happened quite suddenly. The Pueblo Indians left, leaving behind many personal items too heavy to carry—things like large ceramic pots exquisitely decorated in the classic black-on-white pattern common to the time, and sashes woven of animal hair. And they left their city in the cliffs.

What they left behind has been uncovered by archeologists. Why they left is still buried in time somewhere. There is evidence of poor farming techniques which would have robbed the soil of its productivity. There is evidence of increased warfare with a nomadic tribe which swept in and out of the region. Whatever the forces which drove them to abandon their city, scientists agree that the situation was worsened by a severe drought which sucked the area dry between 1270 and the turn of the century. Studies of tree ring data show that the worse of the drought occurred in the years 1273 to 1287. In those years, the Mancos and its tributary streams must have run only with dust clouds, leaving the elaborate irrigation system useless. The same river which brought life to the region slowly dried it up.

By the early 1300s when the rains returned and the Mancos flowed again, the city was empty and would stand silent until the day a wandering cowboy rode up to the edge and peered down through a crack in time some six hundred years later.

Even today the Mancos is unpredictable, subject to flash floods when the dark thunderheads gather over the mesa and drying to a trickle when the heat lays on the land like smoke. It drains a big and rugged country. From the tops of nearby ridges the La Sal Mountains can be seen, shining with the snowcaps of spring ninety-five miles distant, on days when the smog from powerplants in the Four Corners region is light. The views are so long that the mesa was chosen as the site for an experimental relay station in the 1800s for messages flashed from Fort Lewis to troops in the fields.

This is a tough place for a river, a country where the taste of dust hangs on every breeze and each new rain is like a reprieve. For those who live there today, the dry leaves of cottonwoods along the river rattle in the wind and bring back memories of long, dry spells that can hover over the Mancos like vultures and change life in this fragile desert valley forever.

North Platte River

Source: North Park near Walden, Colorado.
Mouth: Platte River, North Platte, Nebraska.
Length: 618 miles (25 miles in Colorado).
Counties: Jackson.
Cities: Cowdry, Walden.
Access: Colorado 125, county and forest roads.

It has been much maligned. The Platte River, in its lower sections, was an important part of the

Spruce Tree House at sunset, Mesa Verde National Park. More than a thousand years ago, the Pueblo Indians depended on the Mancos River for life—when the river failed, only ruins were left behind. TOM ALGIRE.

Called "Bull Pen" by the Indians for the huge herds of buffalo and elk which once roamed there, North Park and the headwaters of the North Platte River now accommodate grazing cattle. LARRY ULRICH.

Oregon Trail leading to South Pass in Wyoming. Many travelers, both famous and faceless, trekked along it. Each one, it seemed, had to get his two-cents-worth in.

Washington Irving, author of such classic tales as *Rip Van Winkle* and *The Legend of Sleepy Hollow*, called the Platte "a magnificent and useless river." Samuel Clemens, known to the world as Mark Twain, commented that if the Platte were somehow turned up on edge, it just might make a respectable river. Others, dealing with the quicksand and mudholes of the lower Platte, dubbed it "a mile wide and an inch deep" or said it was too shallow to float, too deep to ford, and too wide to bridge. Even its early name, Riviere de Plat, meaning "Flat River" in French, reveals the river's deficiencies.

But all of this bears little resemblance to the North Platte in its upper reaches of Colorado and Wyoming. Starting in a tapestry of small, sky-blue creeks in North Park, last of the four great high country parks to be discovered in Colorado, the North Platte is born as beautiful a mountain river as can be found in the Rockies. Its beauty won the praise of none other than the "Pathfinder" himself, John Fremont, whose eyes beheld many of the spectacular scenic wonders of the West during his five historic expeditions.

Today, running the whitewater rapids of Northgate Canyon where the river drops 470 feet in eighteen miles, hundreds of boaters each year see that the Platte is not all flatwater and sandbars. Hiking the many trails in both the Routt National Forest of Colorado and the Medicine Bow National Forest of Wyoming, the tree-lined paths put any thought of mudholes and gravel bars far behind. Perhaps if the North Platte had been better known to the early travelers, the comments would have carried a different tune.

But it wasn't well-known country. North, or "New," Park (as it was called by the mountain men and trappers since it was discovered after South and Middle Parks had already been trapped and hunted out) was a kind of no-man's-land. Ringed by the peaks of the Medicine Bow, Never Summers, Rabbit Ears, and Park ranges, it was tempting but dangerous country. The danger came from warring bands of Shoshone, Cheyenne, Arapaho, Ute, and Crow Indians which often converged on what to them was "Bull Pen," so named for the vast number of buffalo once found there. Consequently, the Oregon Trail left the North Platte near the confluence of the Sweetwater bound for South Pass, and most of the trappers stayed clear—but not Jacques Laramie.

In 1820, despite warnings from knowledgeable men of the mountains like "Broken Hand" Fitzpatrick, Laramie lit out to try his luck in the new territory, relying on his wits, and what he considered his friendly relationship with the tribes, to hold onto his scalp.

Things must have been good for a while, alone in a land where there were almost as many beaver as stars and hunting was as easy as picking flowers. Laramie had time to build himself a cabin with all of those untouched mountains as a backdrop. It was a mountain man's dream, and Laramie planned to stay.

But the dream turned to a nightmare. When their partner didn't show up for the annual fall

The Piedra River may be little more than a trickle in dry years or, with enough snowmelt, it may charge wildly through winding valleys all the way to its confluence with the San Juan. JEFF RENNICKE.

rendezvous, his friends came looking. They found him in the cabin he had built in that mountain paradise, dead at the hands of Indians. Today, the name of the mountain man who, at least for a time, found his place is borne by the Laramie River, one of the North Platte's major tributaries; by a city in Wyoming; and by the geologic event, the Laramide Revolution, which pushed to the sky the very mountains the trapper loved so well.

The place where Laramie found his paradise is now mostly national forest and BLM land open to all, It is a land of high parks, jagged peaks, and thick timber. The timber in places is so tall and thick that one homesick settler named a tributary of the North Platte the Michigan River because the timber stands reminded him of the great stands of trees in his home state.

The Encampment River, another of the North Platte's tributaries, was once the site of great timber camps, where trees were felled near the water's edge and then floated downstream with the spring floodwaters. The cutters, called hacks, and the logs, called ties, rode down the river with the snowmelt, guided by rivermen with long poles risking life and limb to keep the logs from forming a logjam, appropriately called an "embarrass" by the Frenchmen who often worked the drives.

The sight of thousands of "ties" plummeting down the raging Encampment and the drivers riding them like wooden broncos or racing along on shore must have been one of the most exciting sights ever seen on a Colorado river. But the log runs are a thing of the past now. Only the waters of the winter snows run in the spring run-off. Much of the upper Encampment is today surrounded by the boundaries of the Mount Zirkel Wilderness Area and the Davis Peak Wilderness Study Area. The quiet has returned to Encampment.

All of this—the mountain peaks, the buffalo, the thick timber, the clear waters—was missed by Irving and Twain. The early settlers never knew of the deep gorge near its headwaters where the "flat" river was turned on edge through a canyon called Northgate. They never imagined the hillsides full of wildflowers or the silhouettes of the mountain peaks at sunset.

Jacques Laramie knew, for that one idyllic summer he spent in his paradise. Fremont knew. And now a growing number of hikers, campers, and boaters are discovering that the North Platte is even more than the "respectable river" that Mark Twain envisioned.

Piedra River

Source: South River Peak in Weminuche Wilderness.
Mouth: San Juan River at Arboles, Colorado.
Length: 60 miles.
Counties: Archuleta, Hinsdale, Mineral.
Cities: Chimney Rock.
Access: U.S. 160, Colorado 151, county and local roads.

Seeing the lower Piedra River for the first time under the dreary November skies of a low-water year, there is little confusion over how the river got its name. The riverbed in such years becomes a hiking trail, leading from rock to rock, over boulders the size of bears that are cold to the touch and the color of starless skies. The Piedra seems all walls, an alley into the wilderness, a river of rocks.

The Dominguez and Escalante Expedition came this way in 1776, more than two hundred years ago, and to their eyes, too, it was a Rio de la Piedra Parada, a "River of the Rock Wall." Those walls are made up of a series of canyons like First Box and Second Box canyons with their inner gorges carved a thousand feet deep into the Paleozoic rock layers. This hard rock means the river cuts down faster than it can widen itself, forming these narrow and shadow-filled canyons. Those slivers of rock which do shave off the canyon walls are often too big to be moved by the river and so clog the channel, eroding very slowing and creating a river of rocks.

The clogged channel also causes the river to roar with the spring run-off. Even in the dark, dry days of November, the rocks in the channel are river-worn and smooth from the waters that have been pouring over them each spring. The run-off, which usually begins with the warming sun of May, provides enough water for the Piedra to be popular with skilled kayakers. Paddlers also run the lower stretches, where the canyon changes and becomes lighter and less steep, but the challenges of the Box Canyons draw the most attention of paddlers.

In 1979, a large rockslide crumbled down the right wall of First Box Canyon. The fan of the slide pushed the river even more tightly against the left wall, creating just a small slot wide enough to paddle only in certain water conditions. The tight slot has been appropriately named the Eye of the Needle.

Like the Los Pinos to the west, the Piedra, in its upper reaches, is a wilderness river. Beginning

For thousands of years, the water of rivers has meant life to various cultures in Colorado. One of the earliest known inhabitants of these river valleys were the Anasazi, "the Ancient Ones," who vanished about A.D. 1200, leaving only the small rock structures where they stored grain along Colorado's rivers. JEFF RENNICKE.

within the Weminuche Wilderness, it flows down through an area of 41,500 acres known as the Piedra Wilderness Study Area, a tract now under consideration for inclusion in the National Wilderness Preservation System. It then flows into lands belonging to the Ute Indians. The water is clear and provides a habitat for ten species of fish, including the endangered Colorado cutthroat trout, as well as rainbow and brown trout. Late fall brings the kokanee salmon migrating into its lower reaches from the Navajo reservoir where they have been introduced.

Also like the Los Pinos, the Piedra has been

The Cache la Poudre River, the last free-flowing river on the Front Range, has been proposed both as the site for a double-dam system and for protection under the Wild and Scenic Rivers Act.
TOM MYERS/TOM STACK AND ASSOCIATES.

studied for possible inclusion in the National Wild and Scenic Rivers System. A 29.4-mile section near its headwaters has qualified for protection under both the "scenic" and "wild" categories. Although there are substantial timber reserves along its banks, the rugged terrain and steep slopes have prevented economic harvest of the timber using conventional logging techniques. So the corridor of the river remains as wild and untouched as the river itself.

Downstream, there is another kind of wildness. Near Chimney Rock in Archuleta County lies the silent ruins of a mysterious culture known only as the Anasazi, the Ancient Ones. Like the ruins along the Mancos River at Mesa Verde, many of the more than one hundred structures along the Piedra date from A.D. 900 to 1200 and were abandoned during a period of prolonged drought. Archeological diggings, begun in 1921, have revealed only bits and pieces about the Anasazi. As many as two thousand people lived at the Chimney Rock site, which was added to the National Register of Historic Places in 1970.

Rivers like the Piedra have their source in the sky and the whims of weather, and on such a day it is easy to imagine what name the Anasazi might have had for the Piedra when the rains stopped coming. Then, too, it must have been a river of rocks.

Poudre River

Source: Poudre Lake in Rocky Mountain National Park.
Mouth: South Platte River at Greeley, Colorado.
Length: 75 miles.
Counties: Larimer, Weld.
Cities: Poudre Park, Laporte, Fort Collins, Greeley.
Access: U.S. 287, Colorado 14.

The Poudre comes out of the mountains like a short fuse, then fizzles as it winds out to meet the Plains near Pleasant Valley, content to meander a bit before meeting the South Platte near Greeley. This profile—the short, steep rush followed by the spilling out to the plains—is typical of Front Range rivers and has played a large part in determining this river's past. It will, undoubtedly, play a part in its future as well.

The abruptness of the river's canyon walls made the Poudre an early obstacle to travel along the Front Range. In those days, the traveler looking to get to the other side of the mountains had two choices—the Sante Fe Trail far to the south where the Rockies are not as high, or the Oregon Trail through the broad valley of South Pass in Wyoming. What lay between the two, places like the Poudre River Valley, was a kind of no-man's-land. North-south travel was made difficult by the prospect of having to find a fording place, and east-west travel meant looking for a route up the steep canyons.

The more immediate problem, of getting across the river rather than up it, was solved with a bridge constructed near Laporte at the canyon's mouth. Laporte was an early outpost for the Overland Stage and the first county seat of Larimer County. The bridge, and another one built on the Big Thompson nearby, was a toll bridge and collected payment from the hundreds of wagons which crossed it daily during the height of settlement.

Blazing a trail up through the canyon took a bit more trial and error. William Ashely, the fur company founder, pushed a trail through in 1824 on his way to one of the famous rendezvous at Brown's Park on the Green River. The trail was altered and improved by a supply train in 1840 headed for Fort Davey Crockett, which followed the basic route of today's U.S. 287. Later, part of that route became known as the Cherokee Trail when a group of Indian prospectors was led by Lewis Evans through the Poudre Canyon in 1849 in search of gold.

It was on some early trail-blazing trip that the river came away with its name, formally the Cache la Poudre. A story in the February 8, 1883 edition of the *Fort Collins Courier* tells the tale of one Antoinne Janis, the first white settler of Larimer County, who claimed to have been present on the fateful trip which earned the river such an unusual name.

According to the story, Janis and his father were part of an American Fur Company supply train on its way from St. Louis to the Green River in November, 1836. A heavy, wet snowfall lasting several days mired the party in its tracks. After the snow subsided and settled a bit, making progress possible, the order was given to lighten the load of the wagons to keep them from bogging down. The party buried expendable goods, then started a brush fire on top of the site and drove wagons over it to make it invisible to the roving eyes of Indians, animals, and other wagon trains.

The party continued to Green River and, later, several men returned to the site and recovered the cached goods which included several hundred pounds of gunpowder. And, so the story goes, the river flowing through that canyon became the "Cache of the Powder," or Cache la Poudre River.

Although it makes an interesting tale, like so many western stories it is best told around a campfire far from a library or after a few rounds at the bar. Actually, the first recorded use of the name Cache la Poudre shows up in the journals of Captain John Gantt of the Dodge Expedition in 1835, a year before the stashing of the powder in the story of Antoinne Janis. That, and the fact that in another context Janis admits seeing Colorado for the first time in the mid-1840s, make it likely just another tall tale. But too much truth can spoil a good story and, despite the facts, the Daughters of the American Revolution erected a monument on the supposed spot of the cache near Bellvue in 1910, and the river still wears its strange French name.

A digging of another, more scientific sort has uncovered other facts about the long history of man in the Poudre River Valley. At a place known as the Lindenmeier Ranch Site on the banks of Box Elder Creek, a tributary to the Poudre, evidence has been uncovered which points to habitation of the area as distant as ten thousand years ago. Discovered by two amateur archeologists in 1924, the site has given up spearheads, hide-scraping tools, and even one projectile point lodged in the bones of an extinct species of bison. The discovery and carbon-dating of the artifacts has pushed back the estimated date of Folsom Man's appearance on the North American continent and established the Lindenmeier Ranch Site as the oldest dwelling site yet found for Folsom man.

Modern man has settled in the Poudre River Valley and the Front Range of Colorado in increasing numbers. Already 80 percent of the state's population has settled on the Front Range, and the majority of the residents live in what is becoming one large city stretching from Fort Collins to Pueblo. Unfortunately for the Front Range, a majority of the water in the state is found on the Western Range. Transmountain diversions, such as the one which enters the Poudre River near Poudre Falls bringing water from the Laramie River, have been the source of much of the water for this growth. The Laramie-Poudre Tunnel is 7.5 feet wide and 9.5 feet high and nearly 2 miles long. Each year since its completion in 1911, it has brought the Poudre River 16,000 acre-feet of water for use in irrigation and urban supplies.

As the Denver-Metro area continues its rapid growth and transmountain diversions become more a source of controversy than a new source of water, demands on rivers like the Poudre grow stronger. Such demands include water supplies for irrigation, municipal, and industrial purposes to feed the continued growth. A series of dams and reservoirs have been proposed on the Poudre, including the large Grey Mountain and Idylwilde Dams.

Yet water is not the only demand. In a growing urban setting, there is also an increased demand for recreation and natural settings. The Poudre is the only one of the Front Range rivers which has been studied and given a recommendation for inclusion in the National Wild and Scenic Rivers System. It is the last wild river in the Denver-Metro area. A resource such as the recreational opportunities it offers anglers, boaters, campers, hikers, picnickers, and sightseers can be valuable to the quality of life in an urban area.

As with any limited resource, the demands may be more than the Poudre can meet. Can it be both a wild river and provide the water resources for a growing community? As the pressure for an answer rises, the river that once hid gunpowder may itself become a powderkeg of controversy on the Front Range.

Purgatoire River

Source: Culebra Mountains, San Isabel National Forest.
Mouth: Arkansas River near Las Animas, Colorado.
Length: 150 miles.
Counties: Las Animas, Otero, Bent.
Cities: Weston, Sarcillo, Trinidad.
Access: U.S. 160, Colorado 12, Colorado 109, county roads.

The Purgatoire River can run deep and strong or not at all. When rains fall, it can turn a quarter-mile wide in places. In the heat that can stick to you like old grease, it is possible to, as the locals say, "walk across the 'Picketwire' and not get the bottoms of your boots wet." Flash floods are not uncommon, nor is dust.

With no major natural lakes on its 150-mile course and only the Trinidad Reservoir which catches water to be used by the city of Trinidad, the Purgatoire follows the natural rhythms of the weather, taking its source from thin air.

Trinidad State Recreation Area has picnic grounds and places for people to relax at the reservoir formed by the Purgatoire. Fewer, but still some, come to the river as it flows along the eastern border of the Comanche National Grassland. For much of the rest of its way to the Arkansas River, however, the Purgatoire flows through private ranches and homesteads with few roads or trails for public access. Those stretches of the river are less familiar.

For instance, grizzlies once haunted the dusty,

The Purgatoire River, shown here near Higby, has spawned, with the help of irrigation, a rich agricultural valley. KAHNWEILER/JOHNSON.

rock-filled canyons of the Purgatoire—as the great bears did almost all of Colorado. In timbered valleys of the high country, perhaps, a grizzly track would have seemed more natural. But it must have been especially unnerving to find clawed tracks as big as a skillet in the mud along the banks of the Purgatoire.

It certainly surprised and unnerved the men of the 1821 Jacob Fowler Expedition. On their way from Fort Smith in Arkansas to New Mexico, the party camped near the mouth of the Purgatoire River near the present-day town of Las Animas in Bent County on November 13, 1821. Two of the party's scouts surprised a "White Bare" in a thicket along the river. One of the scouts, Lewis Dawson, with only a knife to defend himself, was charged by what the journal-keeper called the "despetert anemel."

Infuriated by one of the camp dogs, the bear mauled Dawson while the rifle of the other scout misfired not once but *three* times before the bear was finally chased off by the dog. Dawson was left with injuries too severe for the scant medical supplies carried by the expedition, and he died three days later, becoming the first known white man to die and be buried on Colorado soil.

Death and lost souls are woven deeply into the history of the Purgatoire River, even accounting for its name. Locals call it the "Picketwire" but its full, and very formal-sounding, name is El Rio de las Animas Perididas en Purgatorio—the River of the Souls Lost in Purgatory.

Apparently early in the West's history, a group of soldiers set out from Mexico in search of the fabled golden cities which were said to glimmer from somewhere in this unknown region. Like most expeditions of the time, it had an army officer in charge, several subordinate officers, and at least one representative from the church. Far from Mexico, in some unrecorded scuffle, the commander was killed by one of his charges who then assumed command of the expedition. Under the less-than-holy circumstances, the priests and many of the civilians refused to go on under the command of a murderer, and turned back, leaving the expedition undermanned and without divine guidance.

The priests and the others arrived safely back in Mexico. The expedition was never heard from again. Later expeditions were to find rusty armor and weapons along the banks of a nameless river in southeastern Colorado and hear stories among the Indians. They even met one Cheyenne who called himself "Iron Shirt" and wore a chain mail shirt similar to the ones worn by the lost expedition. He claimed to have forgotten where he got the armored shirt and no answers were ever found. Assuming the whole party met its end on the shores of the river without the benefit of a proper send-off from the church, the river was given its long and sorrowful name, today shortened to just Purgatoire.

In light of this ungodly event that led to the river's name, it may seem a bit ironic that one of the canyons upstream from Las Animas is becoming known as another Garden of the Gods for its unusual rock formations similar to those found at the real Garden of the Gods near Pike's Peak. The canyon is not well-known, lying as it does on mostly private lands and along an unnavigable river. For now, whatever rock formations there may be are known best, just as this entire river, to the dry prairie winds and the few landowners who live along its dusty banks.

Rio Grande River

Source: Stoney Pass, San Juan Mountains.
Mouth: Gulf of Mexico, Brownsville, Texas.
Length: 1,887 miles (180 miles in Colorado).
Counties: San Juan, Hinsdale, Mineral, Rio Grande, Alamosa, Conejos, Costilla.
Cities: Crede, Del Norte, Monte Vista, Alamosa.
Access: U.S. 160, Colorado 149, Colorado 142.

It is hard to hide a major river. Tucked between the San Juan Mountains to the west and the Sangre de Cristo Mountains to the east, the Rio Grande lies in the most isolated of the four major river basins in Colorado. Although the Rio Grande drains 10 percent of Colorado's land mass, only 2 percent of the state's population lives in the drainage. Half of the basin within the state is publicly owned—BLM or national forests. Here, in these deep forests and high peaks, the second-longest river in the continental United States gets its start. Only the Mississippi is longer.

Like the Colorado River, which comes to its greatest fame far downstream from the Colorado border, the Rio Grande is a much more widely known river on its lower reaches. There, it passes through the scenic Taos Box of New Mexico and the Big Bend National Park in Texas, and acts as the official border between the U.S. and Mexico. This is the section of the powerful river that the words "Rio Grande" bring to mind. This is the section that earns the title "dustiest river in the world" for its deep sediment load, once measured to carry enough silt to bury thirty-seven square miles of land a full foot deep in muck every year. In its lower reaches, where it is known as the "most dangerous river in the world" for its flash

The Rio Grande, cutting into the Taos Plateau on the Colorado-New Mexico border, carries a sediment load measured downstream at 23,750 acre-feet of silt annually.
PAUL LOGSDON.

floods, the Rio Grande is the "El Rio Grande del Norte," the Great River of the North.

Upstream, in Colorado, the Rio Grande is smaller, more personal, and not so muddy. Here, anglers know it for its deep, clear pools which provide habitat for large brown trout in the 22.5-mile section designated as a gold medal fishery. Boaters know it for its short, narrow canyons which provide scenery and whitewater challenges.

A more quiet stretch is known across the country for the annual Wagon Wheel Gap Raft Race held each June. The race features an amateur class just for the fun and sun of it and a professional race which offers top prize money and draws competitors from all across the West. The annual event has become one of the most popular of its kind in the state and is playing a big part in bringing the Rio Grande out of hiding.

Being out of the limelight is not new to the Rio Grande in Colorado. During the rush for gold and silver at Silverton on the Animas River, the Rio

Left: In its lower reaches it may be the "dustiest river in the world," but here in the mountains near its headwaters the Rio Grande plays past the aspen like a cool breeze.
DAVID MUENCH.

Below: The Conejos defines a meandering river. Called "Rabbit" by the Spanish, it twists and turns east out of marshlands in the San Juan Mountains to meet the Rio Grande in the San Luis Valley. PAUL LOGSDON.

Grande became just a pathway for prospectors and their dreams and supplies. The hard-working Otto Mears, who pioneered trails all through the San Juan Mountains, drove a trail up and over Stoney Pass near the headwaters of the Rio Grande to Silverton just eight miles west. It was part of the supply trail which became the life line for booming but isolated towns like Silverton and Durango. Everything the men and the mines needed came from Pueblo over the Sangre de Cristo Mountains, across the San Luis Valley, up again following the Rio Grande, over Stoney Pass, and back down to Silverton—a distance of 250 miles.

At that time, Pueblo was the nearest thing to civilization between the Front Range and the San Juan Mountains. This was wild mountain country, and to some extent it still is. The river flows through the Rio Grande National Forest and the northern section of the Weminuche Wilderness Area. Just fifteen miles from its headwaters, near the peaks of the Rio Grande Pyramids, a government trapper shot what was thought to be the last Colorado grizzly, in 1952. Although there were other sightings reported in subsequent years along the Rio Grande River, the trail of the Colorado grizzly seemed to be getting cold until September 23, 1979, when a lone female grizzly seriously injured a hunting guide near the headwaters of the Navajo River just over the Continental Divide from the Rio Grande. The guide killed the bear with a hand-held arrow.

No grizzly has been confirmed since the 1979 incident. But if Colorado does still have any grizzly bears, they probably haunt the wild reaches between the San Juan River and the Rio Grande.

A more official recognition of the wildness left in the Rio Grande Basin is the recommendation by the U.S. Forest Service that one of its main tributaries, the Conejos River, be added to the National Wild and Scenic Rivers System. Thirty-two miles of this little mountain river draining from the 13,172-foot Conejos Peak have been deemed eligible for protection. The Conejos offers some of the best wilderness trout fishing in the state and, at periods of suitable flow, is a popular expert kayak run. More than that, the Conejos provides wildlife habitat to such species as the bald eagle (once nearly extinct in the Rio Grande Valley), the endangered peregrine falcon, and bighorn sheep. The Conejos, along with the Rio Chama, are the Rio Grande's main tributaries.

It is hard to hide a major river, and the Rio Grande has not gone totally unnoticed. In its lower reaches, "Rio Grande" is written bold and broad across the maps and the lives of the regions through which it flows. In Colorado, it carries a more subtle touch, tucked deeply away in a back pocket of the Rocky Mountains.

Roaring Fork River

Source: Independence Lake at Independence Pass.
Mouth: Colorado River at Glenwood Springs, Colorado.
Length: 75 miles.
Counties: Pitkin, Eagle, Garfield.
Cities: Aspen, Basalt, Carbondale, Glenwood Springs.
Access: Colorado 82.

There would be arguments from Telluride on the San Miguel-Dolores drainage, and from the people at Silverton and Durango on the Animas, and from just about anywhere else where people have grown to love their river, but the Roaring Fork River Valley may be the most beautiful watershed in the state.

Consider the facts. The Roaring Fork River rises at Independence Lake deep in the tallest mountain range in Colorado, the Sawatch Range, which contains thirteen of the fifty-three peaks in the state over 14,000 feet including Mount Massive (14,421 feet) and the monarch of Colorado mountains, Mount Elbert (14,433 feet). As it flows by the Forest Service campgrounds near Independence Lake, the 75,000-acre Hunter-Fryingpan Wilderness Area looms above its right shore and it begins to gather the clear waters of creeks like Maroon, Capitol, and Snowmass flowing down to it from the mountains of Maroon Bells-Snowmass Wilderness, which covers 175,000 acres of some of the best scenery in Rocky Mountains. From above and below, it takes on water from Mount Sopris and Castle Peak which stand north and south of the Maroon Bells-Snowmass Wilderness Area, respectively.

At the town of Basalt, it takes on its first of two major tributaries, the Fryingpan River. The Fryingpan Lakes in the White River National Forest give birth to this tributary of the Roaring Fork. Its name comes from a tale in which a trapper left a

Despite its name, the Roaring Fork settles down a bit as it drops from the mountains of the Elk Range. Here, near Aspen, several kayak schools offer classes in river running. KAHNWEILER/JOHNSON.

frying pan in the notch of a tree to mark the spot where his partner, who was injured in an attack by Indians, lay dying as he went for help. The trick worked and he soon returned to the spot of the frying pan with assistance, but too late. His partner was dead.

Through most of its course, the Fryingpan flows over a beautiful rock formation known as the Red Beds. These rocks were washed from the flanks of

Left: Flowing over formations of limestone and marble as it leaves the Maroon-Bells Wilderness Area, the Crystal River picks up little sediment, remaining as clear as the melted ice from which it runs. The Roaring Fork with its two major tributaries, the Crystal and Fryingpan rivers, may form the most beautiful drainage in Colorado.
KAHNWEILER/JOHNSON.

Below: Knee-deep in an angler's paradise, one of the 700,000 anglers who purchase a Colorado fishing license each year tests an inviting stretch of wilderness trout water, where the scenery is as important as the fish.
KAHNWEILER/JOHNSON.

the Uncompahgre Mountains of long ago and now wrap the river in the colors of sunset.

Farther downstream, the Roaring Fork picks up a river flowing the color of mountain air and appropriately named the Crystal. Flowing on limestone formations which do not cloud the waters with sediment, the river flows true to its name from the base of Purple Mountain in the Elk Range. If its own beauty weren't enough, there is a quarry near the town of Marble where heat, pressure, and water have produced veins of frost-colored marble. Named Yule Marble for a nearby creek, the rock is almost pure white or veined with streaks of light brown. Marble from the Crystal River has been used in the Lincoln Memorial, the Denver Post Office building, and is so abundant that many of the rickety log cabins along the river have pure marble doorsteps. One enormous piece, measuring 14'x7'x6' and weighing 56 tons, was quarried for the Tomb of the Unknown Soldier at Arlington National Cemetery.

All three of the rivers in this watershed are dotted with places to rent cabins, picnic grounds, campgrounds of the Forest Service, and historic buildings. The ridges and peaks between them have a web of hiking and horseback trails, and up and down the rivers, deep, blue holes lure anglers. At the center of it all is the town of Aspen, famous for its world-class skiing and mountain views.

As the Roaring Fork River nears the Colorado at Glenwood Springs, the banks wear a greener tint of pastures peppered with horses. Both the Crystal and Roaring Fork are free-flowing rivers, and so pulse with the heartbeat of melting snows and spring rains.

Beauty is, of course, in the eye of the beholder. Telluride, Silverton, and other mountain towns have their arguments—and their own beauty, to be sure. Still, when the moon rises full and the color of Yule Creek marble over the valley of Roaring Fork, it is worth catching an eddy to stop and watch the night settle on the most beautiful watershed in the state.

St. Charles River

Source: Lake San Isabel in Wet Mountains.
Mouth: Arkansas River near Devine, Colorado.
Length: 55 miles.
Counties: Pueblo, Custer.
Cities: San Isabel, Avondale.
Access: Interstate 25, county and local roads.

The two ranges of mountains come out of the tablelands of southeastern Colorado like a pair of breaking waves—the Sangre de Cristos and Wet Mountains. Both have their high peaks—Greenhorn Mountain (12,349 feet) in the Wet Mountains and 14,294-foot Crestone Peak in the Sangre de Cristo Range. From the Sangres comes the Huerfano River. Out of the Wet Mountains comes the St. Charles.

The St. Charles has a short life as a mountain

Rivers play host to an abundance of wildlife—some endangered and others, like this raccoon on the Cache la Poudre River near Fort Collins, thriving. GEORGE POST.

river, dropping quickly out onto the plains for most of its life. It meanders across the tablelands, crossing Interstate 25 just south of Pueblo, merging with its main tributary, the Greenhorn River, just east of the highway. It continues on to the Arkansas River, resting at the two St. Charles reservoirs.

The St. Charles River first received attention as a border line. In 1836, after declaring independence from Mexico, the new Republic of Texas threatened to claim much of the land in what is now southeastern Colorado. New Mexico, not wanting the lands to fall into the hands of Texas, gave out huge land grants, one of which was the Virgil-St. Vrain Grant bordered on the south and east by the Purgatoire River, to the north by the Arkansas and on the west by the St. Charles.

Not long after that land dispute was peacefully resolved, the St. Charles was feeding one of its first irrigation ditches. Alexander Hicklin and his wife, the daughter of Charles Bent, then governor of New Mexico, received five thousand acres of the Vigil-St. Vrain grant as a wedding present. Their land was on the Greenhorn River, the largest tributary of the St. Charles. Hicklin dug the ditch to irrigate his homestead and built his abode house nearby, meaning to stay. But Hicklin's early death left only his wife and kids to defend the land against claim jumpers. The son was killed in one such attempt and Mrs. Hicklin died soon after, penniless and without the land irrigated by the Greenhorn River.

"Greenhorn" is a common name in southeastern Colorado, hanging on a city, a mountain, a creek, and a river. It comes from a Comanche war chief said by his enemies, of which there were many, to be as ferocious as a bull elk with its horns still in velvet, a greenhorn. His enemies finally caught up with him in 1779 at the base of a

Fall Creek, a tributary of the San Juan River long before it becomes a desert river, creates a suspended ribbon of spray as it drops over Treasure Falls. SCOTT S. WARREN.

large mountain in the Wet Range. There, the ornery chief met his end, and the mountain at his back when he fell became known as Greenhorn Mountain. The Greenhorn River flows from nearby slopes.

The St. Charles, like the Greenhorn, was better known to early travelers than to today's tourists. The two rivers are not the major routes they once were. Once, they formed a fork of the Cherokee Trail which led Indians and white prospectors from eastern Oklahoma, across the Arkansas River, along the Front Range near Fort Collins, up the canyons of the Poudre to the trail heading over South Pass, and to beyond. It became a well-beaten path and a way station called the Greenhorn Inn served travelers as they passed the junction of the two rivers. One of its weary travelers was Kit Carson, the famous guide and trapper who left his mark clearly on the history of the West and his signature scratched into the rock face of a cliff just north of the St. Charles River at the town of Wetmore. It is one of the few reminders of that route along the St. Charles and the Greenhorn which opened the West.

San Juan River

Source: Wolf Creek Pass in the San Juan Mountains.
Mouth: Colorado River at Lake Powell in southern Utah.
Length: 366 miles (86 miles in Colorado).
Counties: Mineral, Archuleta.
Cities: Pagosa Springs, Trujillo, Juanita.
Access: U.S. 160, county and local roads.

The San Juan Mountains and their place in history have been shaped by these elements—fire, water, gold, and ice. From these elements flows the San Juan River.

It is the master river of the premier mountain range in the West. The San Juans were born in a dramatic display of volcanic activity that rumbled through much of the southern part of the state off and on for several million years. The result is the largest single mountain range in the American section of the Rocky Mountains, encompassing more than ten thousand square miles of deep valleys and high peaks. Within the San Juans are thirteen of the fifty-three mountains in Colorado over 14,000 feet—from the 14,309-foot Uncompahgre Mountain to Sunshine Peak, which just makes it at 14,001 feet.

The glaciers which followed the fire of the volcanoes scoured many valleys in the San Juan Mountains, creating a rugged wilderness terrain that, even today, remains some of the most pristine wildland found in the continental United States. In the San Juan Mountains of Colorado, there are six designated wilderness areas totaling 850,000 acres and six other areas under study which could add another 135,000 acres.

The foothills of the San Juan Mountains have gold. The "mineral belt" of the Rockies, a fifty-mile-wide streak of metallic ores formed during the Tertiary Period, is bordered on the southwest by a part of the San Juan Mountains. Such finds as the silver deposits at Silverton and the gold and silver at Ouray have brought fortune and many hopeful settlers to the region.

And then, there is the water. Besides the river which carries the name of the mountain kingdom, thirteen other rivers have their source in the San Juan Mountains. These include eight tributary rivers of the San Juan; the Uncompahgre River, which flows into the Gunnison; the Dolores and the San Miguel, which empty into the Colorado; tributaries of the Rio Grande such as the Rio Chama and the Conejos; and the Rio Grande itself. The San Juan Mountains are a fountainhead for the rivers of the Southwest.

The main wellspring is the San Juan River itself. Its headwaters are buried in the deep winter snows of Wolf Creek Pass, one of the snowiest spots in the Rockies. In winter, tall trees are buried to their highest branches and snowslides rumble down steep cliffs and are diverted over the highway by huge wooden platforms called snow huts. As the sun warms in spring, the hillsides come alive with the sounds of running water.

The San Juan River spends its youth flowing down through the San Juan National Forest on a route parallel to U.S. 160, tumbling through small, rocky canyons. Not far below, it begins to gather its waters: first the West Fork, then the East Fork and Turkey Creek.

The upper reaches, within easy access of U.S. 160, have become a popular spot for campers in the national forest campgrounds, hikers on the rivers of trails, and picnickers on the many sites just along the river. Kayakers also challenge the rocky canyons and anglers test the waters of the youthful San Juan.

As it nears Pagosa Springs, the San Juan begins to change. It widens, and its color deepens. Born at over 12,000 feet, it drops to half that before reaching the Colorado-New Mexico border.

In its lower reaches, the San Juan loses its mountain spirit and becomes a desert river. This is how the San Juan was first known. Dominguez and Escalante passed this way on their 1776 journey to blaze a route from the churches of New Mexico to new missions in California and named the river after St. John the Baptist. To the Navajo, long before the Spaniards set foot in this territory, the waters flowing from the mountains were called Powuska, "the Mad River," for its fits of flash flooding, drought, and many moods in between.

The Indian name is more descriptive, but has been overlooked.

The Navajo River, which in its headwaters hid the last confirmed grizzly in Colorado, may yet fulfill the state's hope of being grizzly country. The Navajo enters the San Juan after once leaving Colorado at Edith and then swinging back north to

No single species of wildlife is more closely tied to the whitewater of Colorado's mountain rivers than the water ouzel, also called the "dipper."
CHARLES G. SUMMERS JR./AMWEST.

join the master river at Juanita. The Navajo River was included in the beautiful Tierra Amarilla Land Grant and its beauty prompted even the footloose Kit Carson to stop and build a cabin on its shores.

The Navajo Reservoir on the Colorado-New Mexico border takes some of the madness out of the Powuska. The huge manmade lake reroutes the river's waters to irrigate the fields of the San Juan Basin. Both the wild Piedra and equally wild Los Pinos join the San Juan River at Navajo Reservoir.

After leaving Colorado, the San Juan sneaks back into the state for just a short stretch at the Four Corners area, the only spot in the United States where a common point is shared by four state borders. Before it reaches that man-drawn phenomena, it takes in the waters of the Animas and Mancos, which both have headwaters in Colorado. From there the San Juan skirts southern Utah, tangling itself in the famous Goosenecks where the river meanders in the depths of a deep sandstone canyon and reaches the backwash of the Glen Canyon Dam and Lake Powell in the Glen Canyon National Recreation Area. There, the powerful flow of one of Colorado River's largest tributaries is stilled, a rather unceremonious ending to the master river which begins in fire and water, gold and ice far upstream where the San Juan Mountains scrape the sky.

Smokey Hill River

Source: Unnamed ridge in central Cheyenne County.
Mouth: Republican River at Junction City, Kansas.
Length: 510 miles (25 miles in Colorado).
Counties: Cheyenne.
Cities: Cheyenne Wells, Arapaho.
Access: U.S. 40, U.S. 385, county and local roads.

Things are not always as they appear. The dusty, high tablelands of Colorado's eastern plains are drained by a few scattered rivers like the Smokey Hill, which winds all the way to the Republican River in Kansas, and the South Fork of the Republican to the north. These are rivers of dust, "an apology for a river," as Horace Greeley called the South Fork of the Republican on his journey by stagecoach into the settlement which would one day bear his name. Although the South Fork is today regulated by two reservoirs and flows through both the Flagler and Bonny state recreation areas, these two rivers are not typical tourist attractions like some rivers in Colorado. The landscape would not rival the Roaring Fork or the Animas.

Yet, to the eyes of a French Army officer in 1724, the area was "the most beautiful land in the world." A hundred years later, another Army officer, this time an American by the name of James Pattie, found it very much to his liking, although he was lost and mistook the Smokey Hill River for a tributary of the Platte. Despite his loss of bearings, he stayed in the area long enough to kill eight grizzlies and fashion a necklace from their claws in hopes the sight of it would convince the Indians of his skills in battle.

Somewhere out on this barren ridgeland ranged not only grizzlies but, sometime in the much more

Left: The Smokey Hill River, one of the true prairie rivers in Colorado, curls lazily through cactus and sagebrush.
KAHNWEILER/JOHNSON.
Below: Cattle amble along the banks of the South Fork of the Republican River near Bonnie Reservoir.
KAHNWEILER/JOHNSON.

distant past, crocodiles. The Hayden Survey Party of 1870 found the fossilized remains of prehistoric crocodiles in the strata which bound the Smokey Hill River.

Grizzlies, even crocodiles, would have been a welcome sight to those unlucky travelers who, eyeing the shortest distance between two points, left eastern Kansas hellbent on the Rockies. The two points were Leavenworth, Kansas, and Denver, Colorado, and the route became better-known as The Starvation Trail.

It was the straightest route across the plains—which was its only advanatage. The route lacked game, adequate cover from the harsh prairie winds and sun, and way stations for resupply. Despite the fact that it cut across the Smokey Hill River, it lacked reliable sources of water. Soon, the unmarked graves and sun-bloated carcasses of stock which lined the trail were proof of the fact that the fastest route is not always as the crow flies.

The *Rocky Mountain News*, the oldest newspaper in the state, ran the story of two more fortunate travelers who struggled into Denver sunburnt the color of old leather, foot-sore, thirsty, and more dead than alive on May 7, 1859. They told the world the horrors of the Starvation Trail, seeing at least a dozen unburied bodies along the route and surviving for nine days on only the water they could squeeze from an occasional cactus and the sparse meat from "one hawk" they somehow managed to kill.

Despite the stories, several attempts were made to promote the route for the construction of roads and highways to Denver. The proposals did not get far, and today, the Smokey Hill River, once the home of crocodiles and grizzlies and the blind dreams of desperate gold seekers, flows alone across an arid land called "beautiful" by some and "deadly" by others. Things are not always as they appear.

South Platte River

Source: South Park in Pike National Forest.
Mouth: Platte River at North Platte, Nebraska.
Length: 424 miles (360 miles in Colorado).
Counties: Jefferson, Douglas, Arapaho, Denver, Adams, Weld, Washington, Logan, Sedgewick.
Cities: Deckers, Denver, Greeley, Fort Morgan, Sterling, Julesburg.
Access: U.S. 85, U.S. 34, I-76, Colorado 9, county and local roads.

When Major Stephen H. Long wrote in 1820 that the area he had just returned from was "almost wholly unfit for cultivation," he was speaking partly of what is now the sugar beet fields, wheat fields, and pasturelands of northeastern Colorado. When he added, "and of course uninhabitable by a people depending upon agriculture for their subsistence," he had no way of knowing that 150 years later more than two million people would live there, between the peaks named for him and for fellow explorer Zebulon Pike—Long's Peak and Pike's Peak.

Long did not foresee the role of the South Platte River—to him just part of his long route of exploration, but in reality a fountain of life for the Front Range.

The South Platte River drainage takes in 28,584 square miles in Colorado, more than the Colorado River or the Arkansas in the state. Large as it is, it makes up less than 20 percent of the state and yet holds more than 60 percent of its population and the greatest concentration of irrigated lands within the state. More than 2 million people in cities like Denver, Boulder, Fort Collins, Loveland, Fort Morgan, Greeley, and Sterling rely, at least in part, on water flowing down the South Platte.

But the South Platte does not do all this alone. The waters of the South Platte or its tributaries such as the Big Thompson and Poudre are augmented by eighteen different transmountain diversion projects bringing 450,000 acre-feet of water once bound for the Pacific Ocean. The biggest of these, the Colorado-Big Thompson Project, brings 234,000 acre-feet of Colorado River water each year to the basin of the South Platte via the Adams Tunnel and Big Thompson River. A more direct feed comes to the North Fork of the South Platte from Dillon Reservoir on the Blue River, bringing 83,000 acre-feet of water through the Roberts Tunnel each year. Much of the water brought into the basin from across the Divide goes to upstream cities and irrigation projects on the Platte's tributaries, but enough reaches the mainstem of the river to meet demand—so far.

The body of the South Platte itself wears a series of reservoirs which on a map make it look like a string of blue pearls. It has not always been that way. The South Platte, the wild and free-flowing South Platte River of the past, was an important highway for early travel. Until the discovery of gold where Cherry Creek flows into the South Platte, in what is now Denver, the major trails missed the Front Range. The Oregon Trail followed the North Platte to South Pass, and the Sante Fe Trail cut across the southeastern part of the state. The discovery of gold in 1858 also meant the discovery of the South Platte.

The trail to the gold at a new city near the discovery known as Auraria followed the same basic route as today's Interstate 76, which means it followed the South Platte River for most of its route. The gold brought hundreds of thousands of people into the South Platte River valley. Some of them got rich. Some of them left. Many of them hung around and began the cities which would grow on the strength of the river they followed when they came.

Although that river looks tame today, and for most of the time it looked shallow and harmless back then, the South Platte had some spirit in it. Floods of 1864, 1875 and 1878, as well as other frequent minor problems, beat back attempts to bridge the river and kept settlers out of the floodplains. But the big one came in 1965.

By that time, it seemed like the South Platte was under control. Shopping centers and housing developments were springing up with confidence within the floodplain of the river. It was this unconcerned attitude which turned the flood of June 16, 1965, into the most expensive natural disaster in Colorado history.

It began as a clear day in Denver, with some isolated thundershowers building to the south and east. By noon, a tornado had touched down fifteen miles east of town, and thirty roofs had been torn

Besides its role as wellspring for the cities of the Front Range, the South Platte is also an important recreational resource, attracting anglers to sections like Eleven Mile Canyon. STEWART M. GREEN.

Broken by dams and reservoirs, cut by irrigation channels and intake pipes, the South Platte River still has not lost its scenic charm. Out beyond the reaches of Front Range cities, it goes quiet, turning the color of gold coins as the sun sets. DAVID MUENCH.

off buildings in the town of Palmer Lake forty miles to the south and near where the South Platte River makes its great bend to the north towards Denver. Rains started in the canyons of the upper South Platte, and an hour later, severe mudslides and flooding carried away homes and heavy equipment. By 8 p.m. that evening, the flood which had turned the placid South Platte into a torrent reached Englewood on the outskirts of Denver, and the devastation began.

The multi-million dollar shopping centers, office buildings, and residential units which had been built on the floodplain were washed away in an instant. Fortunately, with the help of early warning and good communications, most of the thousands of people who lived in the path of the flood had been safely evacuated. In all, six people lost their lives. The town, on the other hand, was ruined. Total costs of the storm will never be accurately known, but best estimates put the cleanup job and damages at $508 million. The Centennial Day flood of the Big Thompson, by comparison, caused an estimated in $35 million in damages but that disaster took a much greater toll in lives.

Chatfield Dam was constructed soon after the 1965 disaster, in large part to facilitate flood control. Cheeseman Dam was already in place upstream. Then came the twin upstream facilities of Eleven Mile and Antero. Tucked in between them, in 1980, was the Spinney Mountain Project, and construction on the latest project on the South Platte is being completed at the Strontia Springs Dam in Waterton Canyon. A new, and what would be by far the largest, project on the South Platte, has been the focus of an intense battle between the Denver Water Board and a coalition of environmental organizations, sportsman's groups, and others concerned that the South Platte has been over-taxed already. The proposed project, known as Two Forks, is still on the drawing board.

Evidence of the overtaxing of the South Platte is not hard to find. Through the urban sections of the river—in Denver other cities—the South Platte is a defeated river. It has been dammed, diverted, polluted, used as a trash dump, and neglected for decades and the effects have begun to show.

The South Platte is the hardest working river in Colorado. Its waters have been the foundation for the growth of cities far beyond the imagination of the Long Expedition, and even beyond the vision of Emma Burke Conklin who wrote in her book *A Brief History of Logan County*, "Wide but shallow, fierce but fallow, the Platte River is destined to gather the melted mountain snows and carry them to the wilderness which will someday blossom as a rose."

Today, the South Platte waters more sugar beet fields and hay meadows than rose bushes, but the point is clear. This river has spawned a great city at the foot of the Rockies. Gazing down on the lights on the Front Range from the peak that bears his name, Stephen Long would be amazed.

Renaissance of a river

When the 1860 Stephen Long Expedition stopped to rest at the spot where Cherry Creek flows into the South Platte River—where the city of Denver sprawls today—the river that flowed by them was clean and wild. By 1972 and the passage of the Clean Water Act, the South Platte would be listed as one of the most polluted and neglected streams in the nation.

The South Platte of the early 1970s was an eyesore, lined with gutted warehouses and rusted equipment and studded with gutter drains spewing untreated wastes and chemicals directly into the water. The city of Denver had turned its back on the problem.

But with the added incentives of the Clean Water Act, groups like the Colorado Open Space Council, Trout Unlimited, and the League of Women Voters began to push for a cleanup of the South Platte. In 1974, Denver Mayor Bill McNichols authorized the establishment of the Platte River Development Committee and a renaissance was begun.

Today, a ten-mile biking and walking trail follows the river connecting a series of parks that sit where garbage dumps once sat. Warehouses have been restored and turned into shops; murals have been painted on the sides of buildings facing the river. Where chunks of unsightly concrete were once placed to stablize the riverbank, natural boulders have replaced them. On the dangerous roller dams which once prohibited river running, boat chutes have been installed to open the river to kayakers and rafts.

The work is more than a cosmetic facelift. Pollution levels have been drastically cut. Now, night herons fish the rapids. A six-acre section called Habitat Park is being restored to its natural state with the replanting of native grasses and trees and the removal of introduced species. Deer, beaver, geese, ducks, and pheasants have returned to the river and many of the 280 species of birds found in metro Denver can be seen along the Platte.

And people are returning to the South Platte. Any warm summer day, the path is filled with joggers and bikers, the parks dotted with families on cookouts or bird watching. The South Platte is back and this slice of wilderness in the heart of downtown has people wondering how they ever let it slip away in the first place.

White River

Source: Trappers Lake in Flattop Mountains.
Mouth: Green River near Ouray, Utah.
Length: 183 miles (112 miles in Colorado).
Counties: Garfield, Rio Blanco.
Cities: Burford, Meeker, Rangely.
Access: Colorado 64, County 20, local roads.

Oil and water don't mix, but on the White River in northwestern Colorado, the blend is being tried anyway. The White River flows, downstream from the town of Meeker, through some of the richest deposits of oil shale in the world. Oil shale deposits, a kind of petroleum-soaked rock, are estimated to contain greater amounts of oil than all of the known crude oil reserves have ever contained. The oil shale deposits of the Piceance Basin—which includes parts of western Colorado, eastern Utah, and southern Wyoming and parts of which drain into the White River—could yield as much as 1.8 trillion barrels of oil.

Trouble is, it takes water to get the oil out, a lot of water, up to five barrels of water for every barrel of oil produced. The Piceance Basin is high desert country rutted with dry, dusty creekbeds which run only for a few weeks in the wettest of springs. Water is harder to find than the oil. To find it, eyes have turned towards the White River.

Currently, the White River is free-flowing. Its headwaters are at Trappers Lake in the Flattops Wilderness Area, which encompasses 235,000 acres of the Routt National Forest. The Flattops is an historic spot in the annals of the wilderness movement, for it was at the headwaters of the White River where a young "recreational engineer" employed by the U.S. Forest Service named Arthur Carhart got the idea which would eventually lead to the establishment of the National Wilderness Preservation System in 1964.

Carhart, sent by the U.S. Forest Service to Trappers Lake in 1918 under orders to survey the area and submit a design for the construction of summer vacation homes around the lake, was struck by the untouched beauty of the lake and the upper basin of the White River—so struck that he returned to his superiors in Denver and told them the best use of the area laid not in vacation homes but in preservation as "wild lands." Although he didn't receive much encouragement from his bosses, Carhart told another ex-employee of the Forest Service of his experience at the headwaters of the White River, and the concept of wilderness was born. That ex-employee was Aldo Leopold, known today as the "Father of the Land Ethic," and author of the outdoor classic, *A Sand County Almanac*.

After leaving the National Forest, the White River flows through a checkerboard of land-ownership, with BLM lands alternating with privately held ranchland. Here, the White crosses a high, lonely desert. The river parallels Colorado 64 for most of its route, but it is a lonely, quiet stretch of road. Below Meeker, the river slips past the Rio Blanco State Recreation Area, where a few canoeists paddle the river each year, and then down past a small park where a few eyes watch it flow by. But, for the most part, this quiet river slips by without much notice. Unless, that is you work for an oil shale company.

At the height of the oil shale boom which hit the West and Colorado in particular, in the late 1970s and early 1980s, the White River was targeted for a major dam just below the Colorado-Utah border. The White River Dam would provide electrical power and water needed by a nearby oil shale production plant. The result would be the end of the free-flowing White River, a fourteen-mile reservoir which would flood the core of the White River Canyon in Utah, the river's most scenic stretch, and disruption of habitat for species like the endangered Colorado River squawfish and three other native species.

The proposed project was put on hold, and the White River given a reprieve when the bottom fell out of federal funding for oil shale research and production. So the White River still follows its quiet route across the western edge of the state, free-flowing and left alone, at least until a way is found to make oil and water mix.

Yampa River

Source: Derby Peak in the Flattop Mountains.
Mouth: Green River at Echo Park in Dinosaur National Monument.
Length: 170 miles.
Counties: Moffat, Routt.
Cities: Steamboat Springs, Hayden, Craig, Maybell.
Access: U.S. 40, Colorado 131, Dinosaur National Monument roads.

To begin to understand that the Yampa is a free-flowing river, there is no need for maps. Just stand on its banks in mid-June. Pick a spot where the cottonwoods along the bank are leafless, gray, and their trunks as smooth as river-worn rocks from too many spring floods. Stand and watch the river carrying whole trees, tangles of brush and fence posts, the spoils of the upstream war with its banks.

Then, return in the heat of September to the same spot and walk where just a few months ago there was a river. The Yampa is like that, a complex and contradictory river.

It begins high in the snows of the Flattop and Gore mountains and then drops through desert canyons. The land around it wears names like Happy Canyon and Disappointment Draw, Starvation Valley and Thanksgiving Gorge, Dry Woman Canyon and Five Springs Draw. Where it drops out of the mountains and meanders in a wide valley, horses graze in knee-deep grass. Where it meets the Green at Echo Park, there is sage brush and slickrock. It is that kind of river.

The first introduction of the Yampa for most is cresting Rabbit Ears Pass on U.S. 40 and starting down towards Steamboat Springs. There, the Yampa Valley opens up wide and green, and the blue thread of the river meanders below back and forth across the valley floor. Here, the Yampa enriches the fertile valley floor.

Many locals still call this the Bear River. Some early maps also show it by that name, a confusion of the translation of the Ute word *Yampa*. The name derives not from an animal but from a wild herb (*Carcum gairdneri*) commonly dried and used for food and medicine by the Ute Indians.

Ranching is still an important resource for the Yampa River Valley, but other new industries have sprung up since the time when only cattle and the river roamed these parts.

Steamboat Springs has grown into an internationally known resort town with winter skiers drawn from around the world to its good snowpack and powder. When that snow melts in spring and flows down into the Yampa River, it provides for another new and vital industry in the valley, whitewater rafting.

The Yampa's 170-mile course provides a rainbow of boating opportunities. Stretches above and surrounding Steamboat Springs offer gentle, flatwater canoe trips through the wide open valley. Downstream, the river gets pinched into a pair of canyons—Juniper and Cross Mountain. The Juniper and smaller Duffy Canyon upstream offer good canoeing and intermediate kayaking. Cross Mountain Canyon, where the river drops two hundred feet in less than four miles, is for experts only. By far the best known of the boating sections of the Yampa is the forty-six-mile section within the boundaries of Dinosaur National Monument.

By the time it reaches the eastern boundary of the Monument, the Yampa has taken in its two main tributaries. The Elk River, which begins in the Mount Zirkel Wilderness across the Divide from the headwaters of the Encampment, enters just downstream from Steamboat Springs along

The Yampa, pictured here at Harding Hole, is the longest free-flowing river left in the Colorado River Basin.
MELINDA BERGE/ASPEN.

U.S. 40. The other major tributary, the Little Snake River, slaloms its way along the Colorado-Wyoming border, entering and leaving the state three times before finally settling on a southerly course and meeting the Yampa at Lily Park below Cross Mountain Canyon. Both tributaries add substantial amounts of water during spring runoff, and by the time the Yampa reaches the Monument, it flows with gathered strength.

That strength has helped the river carve its lower canyon through an eastern spur of the Uinta Mountains, the largest east-west trending mountain range in the continental United States. The canyon it has cut is as spectacular as any in the West.

The forty-six-mile canyon borders the stretch of the river which has been recommended for inclusion in the National Wild and Scenic Rivers System. It is habitat for species like the endangered peregrine falcon, which nests in the safety of cracks and ledges on the sheer cliff faces. Several research projects into the population and re-introduction of the peregrine are currently underway in Dinosaur National Monument.

The river is also habitat for some less-glamorous species, also threatened with extinction. Four species of fish native to the Colorado River watershed are making their last stand here. One of these is the Colorado River squawfish, which can grow to lengths of six feet and weigh up to eighty pounds and is, strangely enough, a member of the minnow family. Once, the squawfish was abundant enough to support commercial fishing in some lower basin rivers. Its value as a food source would probably be disputed, however, by the head boatman of the Powell Expedition in 1869 who described its taste "like a paper of pins cooked in lard oil." Yet it was not the eating but the disruption of its habitat by the sixty-five dams on the Colorado River Basin rivers which left this species with just one small bend in the Yampa River as known breeding habitat.

The Yampa Canyon has been home to humans as well. At sites on the river it is possible to look into the eyes of another culture through the pictographs and petroglyphs of the Fremont Indians. Shelters and overhangs in the area which served as hunting camps or living quarters for the Fremont Culture have revealed such archeological finds as a ceremonial headdress made of a mule deer hide with the ears intact and a medicine bag containing 350 flicker feathers.

Such artifacts are valuable to science, and at least one resident of these lonely canyons grasped their importance early.

Pat Lynch was the hermit of the Yampa. He lived in several places up and down the canyon from the late 1870s to 1917, exploring, gardening, carving his own petroglyphs in the rock. In one of the shelters he used, he left a poetic hand-written note that read, "If in these caves you shelter take/Plais do them no harm./Lave everything you find around/Hanging up or on the ground."

It is sound advice, for the box canyons and shadow-filled corners of Yampa Canyon may yet give up more of their secrets to science.

Even with all the wildlife and pictographs, the real value of the Yampa is in its wildness, the way it winds like a dropped lariat at Serpentine Bends, where the river travels seven miles to cover a single mile as the crow flies, and then crashes straight through Warm Springs Rapids with a sound like thunder. It is a place where the history is natural history, written on the tree rings of drift logs and the hieroglyphics of heron tracks in the sands where it meets the Green River below the cliffs of Echo Park. It is a wildness that is rare in today's world. The maps say that the Yampa is the longest undammed tributary in the Colorado River Basin, but maps are of little help in coming to understand the Yampa. It is best just to pick a spot on its banks and watch it flow by. The river will reveal its secrets in time, or maybe not. The Yampa is that kind of river.

Chapter four:

Divided waters, divided people

They are sparse, brightly lit rooms, glimmering in chrome and glass. Each wall holds row after row of gauges and meters glowing a metallic green. Panels of digital read-outs flicker with endless strings of numbers. Phones ring, printers chatter quietly off in a corner, and the keys of computer terminals cluck nervously like cracking knuckles.

These are the "mission control" centers of Colorado's rivers: command posts like Power Control Center in Montrose and Lake Estes Power Center in Estes Park. Here, a technologic thumb is held to the pulse of the state's rivers. A twitch of a meter is the stirring of the Gunnison River as it tumbles through the turbines at Morrow Point Dam. A flash of numbers is the Big Thompson as it rises restlessly behind Olympus Dam. With just the flick of a switch, the Green River can be made to rear up and spin the generators at the Flaming Gorge Dam before sliding off towards Lodore Canyon, or the South Platte River can be slowed to a trickle while irrigation waters are diverted.

The sun may still melt the snow and the rains may still raise the creeks, but it is the engineers in these glass rooms across the state who conduct the music to which the rivers of Colorado must now dance.

Beyond the walls of the control rooms, it is spring. The breeze over Palisade Valley carries a faint hint of peaches. The fruit trees are in bloom, each branch swaddled in petals the color of white silk. To the east, beyond the cities of the Front Range, the swells of plowed earth wear the green tint of the early wheat crop, while a Bent County farmer is out walking in fields which will be knee-high in corn by the Fourth of July. To modern eyes, the first stirring of the fields has become a sure sign of spring, as traditional as the splash of wildflowers across a hillside. Yet to the eyes of early explorers like Long and Powell, peach blossoms and cornfields in the Rockies would have seemed like a mirage.

In 1820, when Long traversed the Front Range from Long's Peak to Pike's Peak, his eyes fell on a landscape which bore a "manifest resemblance to the deserts of Siberia." In a report to his superior officers, Long stated that while the area did present a defensible position in the event of military maneuvers, the climate and lack of water made it, in his eyes, "almost wholly unfit for cultivation."

For Powell, seeing Colorado a half-century later and through the eyes of one trained in the sciences, the conclusion was only a little more promising. In addition to his famous river expeditions, Powell was appointed to head an 1878 survey of the West's irrigation potential. In his report, Powell became the first to classify Colorado's climate as "semi arid." Like Long and others before him, Powell emphasized the dryness of the statewide annual average of just eighteen inches of precipitation. He sounded a warning of the limits such a climate posed to the cultivation and settlement of the region. "The West," he said, "is an arid land, hostile to farming, and will never be settled . . . unless the government dams the rivers and saves up the winter and spring run-off in artificial lakes and reservoirs."

But Powell, as the first appointed director of the U.S. Geological Survey, had a vision. Once, he called the West "a region of the wildest desolation." But as a scientist, he was aware of the great steps being made in reclamation, and he saw in Colorado the potential for a place where ". . . man lives in the desert by guiding a river thereon and fertilizing the sands with its waters and the desert is covered with fields and gardens and homes." It was a vision which came to pass. The rivers have, indeed, come to fertilize the sands.

So the desert may bloom

The rivers have been pooled. Within the 1,900 reservoirs that dot Colorado's landscape like giant buckets, 8.85 million acre-feet—equal to one-half of the state's yearly precipitation—can be stored. The reservoirs moderate the spring run-off which rushes from the mountains in a torrent accounting for more than 50 percent of the state's annual flow in the span of just two months, and those same reservoirs hold it against the heat of summer for irrigation.

The rivers have been diverted. Today, the state of Colorado ranks third, behind only California and Texas, in acres of land under irrigation. Four million acres, or 58 percent of all of the state's farmlands, receive water diverted from the state's rivers and streams. The thirsty ditches crisscrossing the orchards and fields are by far the biggest consumers of water in the state, drinking in 79 percent of all the water used annually.

The rivers have been reversed. In sixteen ditches and eight tunnels, 630,000 acre-feet of water, once bound for the Pacific Ocean via the Colorado River and its tributaries, flows backwards and across the Continental Divide towards the Gulf of Mexico in the basins of the Rio Grande, Arkansas, and South Platte rivers. That feat is equivalent to turning a river the size of the Animas about-face and directing its total annual flow over the mountains. The South Platte Basin alone receives 437,000 acre-feet, a flow as large as the White River, to help quench the thirst of the Front Range cities and give the green tint to early wheat.

And the rivers have become the center of controversy. Powell and Long were not wrong. Beneath the green veneer which this manhandling of the state's rivers has spread across the landscape, there still beats the heart of a desert. No amount of dials and switches can change that fact and, although the desert has been made to bloom, the task has spread the rivers thin. Continued increases in population, unchecked pollution, development of water-dependent industries like oil shale, and the lack of statewide planning all threaten to drain the rivers dry.

Proposals to wring even more water from the

The rivers of Colorado can satisfy a multitude of human demands—and they do. However, the rivers have limits and not everybody agrees on what is the best use of the rivers of Colorado. Clockwise from top left: Rafters lurch down the Arkansas.
SPENCER SWANGER/TOM STACK AND ASSOCIATES.

An irrigation channel near Westcliffe serves as the vital link between snows of the peaks and crops of the lowland fields. MICHAEL WEEKS.

From underneath the Continental Divide, over a quarter-million acre-feet of Colorado River water annually explodes into lakes such as Marys Lake, part of the Colorado-Big Thompson Project.
KENT AND DONNA DANNEN.

A string of city parks, 10.5-mile bike trail, and efforts to clean up the polluted water have brought people back to the banks of the South Platte where it flows through metropolitan Denver.
KENT AND DONNA DANNEN.

rivers involve construction of more dams, creation of more reservoirs. But the prime sites for such projects have already been used, and only marginally-efficient spots remain. Programs of conservation, recycling, less water-consumptive agriculture, and strict water regulation and planning can squeeze some additional water out of projects already operating, but capacities are within sight and they are set by water itself.

Through the center of all this controversy flow Colorado's few remaining wild rivers. Only eleven of the state's rivers contain sections which remain wild enough to have been recommended for inclusion in the National Wild and Scenic Rivers System. Efforts to secure congressional approval for this federal protection have been logjammed. Not a single mile of protected river flows within the state of Colorado. Time may be short. Already much of the wildness of the Dolores, one of the eleven rivers recommended for protection, has been tamed by the McPhee Dam. Others—the last wild sections of the Gunnison, the Yampa, the Poudre—face threats which could destroy the very qualities which earned them recommendations for protection. As the noose of limits draws tighter, both the value of protecting these wild rivers and the difficulty in securing that protection grow greater.

Water projects have provided for irrigation, for industry, and for growing cities. But wild rivers provide recreation, wildlife habitat, and fisheries. Water projects have painted the landscape with broad brushstrokes of green—alfalfa, corn, wheat. But wild rivers paint the landscape with colors of a different sort—whitewater, rainbow trout, the sky-blue heron. It is, on both sides, a fine and delicate balance, one which can only be maintained by protecting what rivers still flow free. Despite all the power of technology pent up in those glass rooms filled with switches, there is no switch that can make a river once more flow wild and free.

Tombs and ruins and water law

It is not a new idea. The knowledge required to scratch the landscape, to bring water from where it is to where it isn't, is as old as ancient Babylon. On the tomb of King Hammurabi, buried more than four thousand years ago, the leader is praised for his irrigation systems which "made the water flow in the dry channels and have given an unfailing supply to the people."

A dusty ring of silent rocks on a ridge overlooking the valleys of the Mancos River in Colorado shows that the gift of flowing water was known early here as well. The ruin, known as Mummy Lake, was used by the Pueblo culture around the time of Christ to gather the water of winter run-off and summer rains and distribute it over fields of squash and corn. The site is now a part of Mesa Verde National Park.

When the Europeans came to Colorado, they brought along not only dreams of great cities of gold but also a deep understanding of the ways of irrigation. The first-known attempt at irrigation by settlers came in 1787. In that year, Juan Bapista de Anzi, governor of the Spanish province of New Mexico, signed a treaty with the Comanche. As a part of the agreement, twenty skilled Spanish laborers were sent to a point near the confluence of the Nepesta (now the St. Charles) River and the Arkansas River, where a colony was to be constructed. An irrigation ditch was dug and several buildings erected before a new governor was empowered, the treaty nullified, and the settlement abandoned.

Gauging the flow

Establishing the Prior Appropriation Doctrine and infusing order into the state's water law was only part of the battle. If not for reliable methods of measuring the water parceled out by the hard-fought decrees, shovels and shotguns would still rule the water world. Fortunately, several accurate gauging procedures have been known to science for centuries.

Water has many forms—in the rivers its motion must be measured; in a reservoir, it is quantity which becomes important. To suit these conditons, two systems are used.

Moving water is measured in cubic feet per second, a measurement of the volume of water passing a fixed point in a standard time. One cubic foot per second is equivalent to 7.5 gallons, or 450 gallons per minute. This standard is used to determine the quantity of water which flows through a river's course annually.

A less static measurement is needed for water used for irrigation and stored in reservoirs. That measurement is the acre-foot. An acre-foot is the amount of water needed to flood one acre of land one foot deep, or 325,851 gallons. A single acre-foot is enough water to supply the needs of four families for an entire year, flush 65,170 toilets, brew 2,100 barrels of beer, or fill 25 large swimming pools and several hot tubs.

These two measurements keep the peace in Colorado water law by insuring that the flow of water to industry, agriculture, and municipal users is accurate.

In 1851, a group of Spanish settlers came to Culebra Creek in San Luis Valley, bringing with them a long heritage of farming by irrigation. One of the first acts of the new town, dubbed San Luis, was the cooperative digging of an irrigation ditch from Culebra Creek. With only hand tools and plows pulled by yokes of oxen, the San Luis Peoples Ditch was carved and a year later the waters began to flow. Over 130 years later, the water is still flowing in that same ditch, a tribute to the skill and workmanship of the diggers. In 1890, the District Court of Costilla County awarded the San Luis Peoples Ditch "Priority 1" in the state's water system, making it the oldest recognized water right in Colorado. More important, perhaps,

Keeping the "wild" in wild rivers

Once, and not so long ago, all of the more than three million miles of U.S. rivers ran free. But in 1982, a Nationwide Rivers Inventory conducted by the National Park Service could find only 61,700 miles, or less than 2 percent of all rivers in the lower forty-eight states, which were still wild enough to qualify for study as potential members of the National Wild and Scenic Rivers System. The nation's wild rivers are an endangered species.

Enacted in 1968, the National Wild and Scenic Rivers Act was an effort to keep the wild river from becoming extinct. From the original eight rivers totalling 789 miles protected by passage of the law in 1968, the system has grown to a current total of 7,217 miles of 65 rivers. Yet, even that number represents only 10 percent of the remaining wild rivers in the United States.

Under the federal system, rivers which display "outstandingly remarkable scenic, recreational, geologic, fish and wildlife, historic, cultural, or similar values" can be designated by Congress for protection under one of three categories: Wild, Scenic, or Recreational.

Although no rivers in Colorado have yet been given protection under the system, the state has rivers representative of all three of the categories. A Wild river is exemplified by the upper Encampment, which is free-flowing, accessible only by trail, and representative of what the bill calls the "vestiges of primitive America." A river qualifying under the Scenic classification must also be free-flowing, but may have limited development along its shores, as does the North Fork of the Elk which is followed by Forest Service roads leading to a campground. The third designation, Recreational rivers, is meant to insure that rivers with valuable resources other than wildness are not overlooked. In this category, rivers like the lower Poudre (which is paralleled by Colorado State Highway 14 and checkered with summer cottages) can be protected for their high-quality fishing, recreational, or historic values.

A qualifying river may have more than one designation along its course, but once it has been given congressional approval to join the ranks of the National Wild and Scenic Rivers System it is guarded against any federally-licensed or assisted project which would degrade the values for which it has been protected. Land-use provisions protect the scenic values of the designated rivers and water quality provisions help to insure quality wildlife habitat and fisheries. No damming, diverting, or channelization may occur on designated sections. The river runs free.

The following chart indicates which of Colorado's rivers have been studied for inclusion under the National Wild and Scenic Rivers System and which have been recommended for protection. Currently, not one mile of any Colorado River is protected by the system.

River	Miles Studied	Wild	Scenic	Recreational
Big Thompson	12.6	0	0	0
Colorado	55.7	13	38.7	4
Conejos	45.2	22	0	13.2
Dolores	186.0	33	41	66
Elk	35.3	16.7	12.4	6.2
Encampment	20	20	0	0
Green	36	18	0	18
Gunnison	29	26	0	0
Piedra	53.1	32.5	12.9	5.5
Los Pinos	54	54	0	0
Poudre	74	25	0	44
Yampa	47	47	0	0
Totals	**647.9**	**307.2**	**105**	**156**

the water that flows in the San Luis Peoples Ditch has nourished the people of the oldest continuous town in the state—San Luis.

Corn, beans, and squash—not decrees or appropriations—were on the minds of the people of San Luis when the first turn of the soil for the ditch was made. But the rumbles of the state's first water disputes, which would lead to Colorado's unique system of water law, were not far off. When those disputes came to a head, it was like the sudden collapse of flood gates.

Water was the most quickly developed, excessively apportioned, fought-over and cussed-about resource in the state's history. As gold was being discovered at the confluence of Cherry Creek and the South Platte River in 1859, another more quiet "rush" was also about to begin.

While all the attention was turned to the gold rush, private irrigation ditches were springing up in the 1860s faster than brothels in a boom town. For every prospector who struck it rich in the gold fields, there was a fistful who were striking it rich in the potato fields. One such man was Rufus "Potato" Clark.

Clark was an ex-sailor, ex-prospector, and sometimes-farmer who followed the rush of 1859 to the banks of the South Platte River. There, he plowed some land, dug a ditch for irrigation, and began planting potatoes with an eye towards supplying the mess tents of the mining camps. In 1867, his potatoes were as good as gold and his enterprise made $30,000. Soon he was one of the wealthiest people in early Denver, donating eighty acres of land to Denver Theologic Seminary and even pledging one days's profits from his potatoes to the relief fund for victims of the Great Chicago Fire.

Not all of the early water miners were as generous as Rufus Clark—with their money or with their water. Within three years of the Colorado gold rush, water for irrigation was being siphoned out of all the major rivers in the South Platte Basin, and rivers that were not all that big to begin with were being asked to fuel a boom town. Water, the state was about to learn, could play out quicker than a pair of deuces in a poker game.

Early in the game, "shovel rights" prevailed—he with the biggest shovel got the biggest share of the water. But even as early as 1864, water in rivers like the Purgatoire and the South Platte was being stretched too thin and shovel rights became "shotgun diplomacy."

In an attempt to settle the disputes without shotguns, makeshift water courts were set up. At those early courts, lawyers were barred from the proceedings on the belief that an honest man could argue his own case. To this day, water court is one of the few halls of justice where Colorado citizens still routinely present their own defense, although the laws have become so tangled that few major cases are settled without legal help.

In 1861, the legislature of the newly established Colorado Territory legalized the withdrawal of water for irrigation purposes but stopped short of enforcing any workable system of prioritizing claims. That knot would be left for the next decade, 1870-1880, and the development of a system of water appropriation known as the "Colorado Doctrine." That doctrine, articulated for the first time in Colorado, now serves as a framework for modern water law across the West.

First in time, first in right

As the snows of Christmas Eve, 1869, fell on New York City, something was taking place in that town which would shape Colorado water law. It was a meeting of the committee to form the Union Colony, addressed by both Nathan Meeker, who would later die at the hands of Ute Indians in the 1878 White River Massacre, and Horace Greeley, publisher of the prestigious *New York Tribune*. That night the Union Colony was officially founded, and one year later, after being driven out of the San Luis Valley by a deep snowstorm, the founding settlers laid claim to 72,000 acres at the confluence of the Poudre and South Platte rivers.

Meeker, from his long years as the agriculture editor of Greeley's paper, had written long rows of

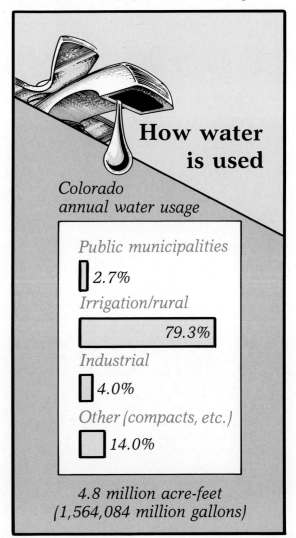

How water is used

Colorado annual water usage

Public municipalities
☐ 2.7%

Irrigation/rural
☐ 79.3%

Industrial
☐ 4.0%

Other (compacts, etc.)
☐ 14.0%

4.8 million acre-feet
(1,564,084 million gallons)

Divided waters, divided people / 95

Like nearly 4 million acres of Colorado farmland which depend upon irrigation, this small farm needs the Dolores River to survive. Irrigation accounts for 79 percent of all water consumed by the state. MARK GAMBA.

words about irrigation. Turning those rows into rows of corn to be watered by irrigation ditches dug by his own hand turned out be a different story. A ditch company was formed at the Union Colony and a fund of $25,000 to cover the construction of four ditches set aside. The Greeley No. 3 ditch was the first to be dug. In just two months a ditch nine miles long, six feet wide and three feet deep had been carved out of the south bank of the Poudre and was bringing water to the crops.

All was not coming up roses, however. Only a small fraction of the projected five thousand acres which was to be irrigated by the project ever saw water and delays and construction errors chewed away at the fund until nearly the entire $25,000 meant for all four ditches had been dried up.

The Greeley No. 2 ditch proved an even more ambitious project and nearly became the death knell of the young colony, which had changed its name to Greeley in honor of its founder. Two thousand acres of crops were confidently planted in anticipation of the water that the new ditch would bring. Nearly all two thousand acres were left shriveling in the sun when delays and errors plagued the project. The construction of the ditch was an exercise in trial and error. Precious time and a lot of money was wasted correcting errors such as the removal of right-angle turns which caused pooling and slowed the flow. When at last water did flow through Greeley No. 2, a much smaller acreage was served and at a much higher cost than had been projected.

Still, good publicity from a fledgling newspaper, the *Rocky Mountain News*, brought a new wave of settlers and colonizers to the region, and the Agriculture Colony was founded on the banks of the Poudre, upstream from Greeley. The founding fathers of the Agriculture Colony also understood the need for irrigation, even if they understood little about the limits of the Poudre River. In 1874, the ditches of this new colony were complete and anxiously awaiting the spring run-off to water crops. It almost never came.

A drought had hovered over the Front Range of Colorado that winter, leaving only a thin snowpack in the hills. With the wide mouths of both the Agriculture Colony ditches and the Greeley ditches waiting downstream, the Poudre River couldn't take it. By the time the new upstream colony had taken its water, the Greeley settlers were left with nothing but dust in their ditches and anger in their throats. An emergency meeting was called in July. Threats were tossed about, fist-fights narrowly avoided, but no agreement was reached. Only a series of heavy rains late in the summer saved the Greeley crops and averted violence. But notice had been served. The rivers were not an endless ribbon of water; something had to be done.

There were as many ideas for how to parcel out the waters of Colorado's rivers as there were rocks in those rivers. The Riparian Doctrine was the law governing the homelands of most of the settlers and still governs much of the United States today. Under the system, a landowner along a river is given water rights as part of his property claim and water is not treated as an entity separate from real estate. In the West, with so many day's rides between rivers and so many weeks between rainstorms, the system fit like another man's boots.

The late-comers, who were forced to settle the upper river valleys, had another idea—rights should go to those who could use the water first, namely those upstream. Since the older and more politically powerful settlements lay on the flatter, more fertile lands in the lower valleys downstream, that idea dried up quicker than an August rainstorm.

The downstream colonies had their own view of how the situation should be handled, a system of "first in time, first in right," and when two citizens from Greeley were appointed to sit on the Irrigation, Agriculture and Manufacturing Committee for the writing of the Colorado Constitution, that system had the inside track.

The Greeley members, whose settlement had been irrigating from the Poudre longer than any large colony on the river, pushed for what they called a "Doctrine of Prior Appropriation." Their plan called for rights to be granted in chronological order dating from the earliest documentation of a "beneficial use." With irrigation the most obvious of these beneficial uses, and with new plans for yet another ditch on the Poudre even farther upstream than the intakes for their own ditches, the Agriculture Colony which once pushed for the rights of upstream use suddenly saw the wisdom of "prior appropriation" and came to its support.

Article XVI, sections 5 and 6 of the Colorado Constitution, adopted in 1876, carved the doctrine of prior appropriation into the law books, and "first in time, first in right" became the foundation of Colorado's distinctive water system. Under the system, priority or "senior" rights are granted the first documented beneficial use on a river. Lesser, or "junior," rights are granted on what water remains based on date of diversion, until all the water in a stream is allocated and the stream dries up—unless the state has been granted an instream flow decree for the protection of wildlife, fisheries, or recreation or unless what water remains in the stream has been promised by a compact to a downstream state. In case of a low-water year, the rights with the most senior of the decrees receive water first, and the intakes of the junior rights are closed off.

Gone are the shotguns and shovel rights, replaced with high-powered lawyers and hills of paper decrees. But the spirit of the West lives on its water. No single topic, not even women or football, can spark a bar fight faster than mention of a water right won or lost.

Men to heed the call

Beneficial use became the rallying cry for ditch-diggers and dam-builders. If, as the Colorado Constitution said, water had no monetary value flowing "wasted" along its course, then the only way to make it turn a buck was to divert it. Until 1969, when recreation was recognized as a beneficial use for water left in-stream, and 1973, when protection of wildlife habitat and fisheries was similarly recognized, the only good water was diverted water. Any project which could bring water to shriveled crops and tongue-dry families was embraced and promoted by the public and the government with flag-waving patriotism. Over no single Colorado project was the flag waved higher than the building of the Gunnison Tunnel.

By the final decade of the ninteenth century, many of the rivers were being tapped for everything they were worth. This included the Uncompahgre River flowing through a valley deep in fertile soil but shallow in water for irrigation. The people of the Uncompahgre Valley could almost hear the music of the key to their prosperity: the waters of the Gunnison River that flowed just sixteen miles distant but was blocked by the high walls of the Black Canyon.

In 1890, a Frenchman by the name of F.C. Lauzon showed that a gravity-fed tunnel could bring salvation to the valley in the form of waters of the Gunnison, but his plan was termed "crazy" and was voted down. For another decade, the people of the Uncompahgre Valley listened at night to sounds of the Gunnison River and their fortunes flowing away downstream.

The whole country felt for the people of the dry valley and the call went out for heroes. In a chapbook written about the tunnel project and the Uncompahgre Valley published in the early 1900s, author Barton Marsh cried: "Where were the men to heed the call and take their lives in their hands by facing the perils of the Black Canyon in the interest of that project which alone could afford relief to the drouth-stricken inhabitants of the valley?"

In 1901, the heroes stepped forward: William Torrence and five others from the nearby towns of Delta and Montrose. With two wooden boats, supplies for a month, and the prayers of all who watched from above, they entered the unknown Black Canyon. Men were stationed at one mile intervals along the lip of the canyon to watch for signs of life in the dark pit below and report the progress of the men to their families and the public. After five long days with no sign of the men, the word was given to erect a mesh net over the mouth of the canyon downstream to seine for their bodies.

The men were battered, beaten, half-drowned, and suffering from what the chapbook described as "that most unpleasant sensation that comes from having water-soaked clothing," but they were all alive. They were finally spotted from above and a general cheer went up as they vanished again into the awesome abyss. For three weeks the men gallantly inched their way among the coal-colored boulders and raging torrents of whitewater, managing to cover only twenty-five miles of their journey before they confronted what seemed like an impassable obstacle. At a place they named Falls of Sorrow, the brave voyagers climbed back out to the world of light and the question of the tunnel remained unanswered—but not for long.

The following year, Torrence was back, this time with two rubber boats, rubber bags for his gear, and a partner by the name of A.L. Fellows, who was an engineer for the newly formed Reclamation Service (later to become the Bureau of Reclamation) which had taken an interest in the project. This time, the two planned to swim the river in the low water of August and complete the survey.

Without incident, the two swam up to the Falls of Sorrow where the other party had been forced the year before to abandon their attempt. Not far below, as described by Marsh, they encountered a falls which spanned the entire river, bound by canyon walls worn as slick as black ice by the water. From the rocks, they could see the tops of trees below the falls but had no way to reckon the extent of the drop or the fury of the river as it dropped out of sight. After hours of searching for an alternative, it was grudgingly agreed that the river posed the only possible path. Fellows went first, leaping into the river's current with a sudden scream, Torrence yelling "goodbye, for he never expected to see him again." Miraculously, Fellows appeared out of the spray and whitewater below, alive, and Torrence had no choice but to jump in and follow.

They were safe, for the moment, but the swim had cost them much of their gear and supplies. Hunger and more cold swims began to take a toll, and just when it looked like they would survive the rapids only to be done in by hunger, Torrence captured what he called a "mountain sheep" with his bare hands and starvation was averted.

In the miles of unknown canyon the two traversed, they were forced to swim seventy-two times in rapids and blind sinks where the river disappeared from the light into churning holes which sucked them in and spit them out downstream like driftlogs. Still, it was engineering and not adventure which coaxed them into the canyon, and ten days later the pair emerged with the verdict. Yes, the tunnel was possible.

After the heroics of the survey, the construction of the tunnel seemed anti-climactic, yet it was not without its moments. Several times, the sharp drill bits tapped into cauldrons of subterranean hot springs, flooding the tunnel in seconds with a wave of scalding water. The drills also poked into underground bubbles of poisonous carbonic gases, filling the tunnels with choking, evil air which could only be vented by drilling an eight-hundred foot shaft straight up. Several workers lost their lives in the construction; many others came, saw, and left. The average worker tenure over the five-year duration of the project was less than one month.

But on July 6, 1909, the two drilling teams met and daylight flooded the tunnel. Then on September 23, 1909, President Taft laid a finger on an electronic switch which unleashed a peal of electronic gongs, telling the world that 200,000 acres of the "Great American Desert" had been reclaimed. Although that 200,000-acre goal would never be reached, the longest tunnel in the world when it was built—a tunnel the *Denver Post* proudly proclaimed as big enough to float a battleship through—brought the sound which the people of the Uncompahgre Valley had waited a decade to hear—running water. The U-shaped tunnel has the capacity to bring a maximum of 1,300 cubic feet per second of water to the 62,000 acres planted in hay, beans, and some of the largest cherry and peach orchards in the central Rockies.

Today, transbasin diversions are commonplace; a flow of 1,050,000 acre-feet a year flows over, under, around, or through the Rockies in Colorado, including the half-million acre-feet of trans-Continental Divide projects. But in 1909, the Gunnison Tunnel was an achievement of good old American sweat and ingenuity. It turned the dry, windy Uncompahgre Valley, in the words of that 1909 chapbook, into "a place sought out by all who love to enjoy the best the earth affords" and a place to find "the ideal man, strong in body, keen of intellect, deep souled, full-hearted and unafraid."

From that point on, the rivers would never be the same.

Sparks across the mountains

The power of water is apparent—in the canyons it has cut, in the booming voice of the rapids it crashes through, in the boulders strewn by flash floods. Projects like the Gunnison Tunnel proved that rivers, like horses, could be broken and set out to pasture. The next step was to lasso wild rivers and bridle their energy for the production of raw power.

The muscle in water comes from its fall, and with many rivers in Colorado dropping thousands of feet from headwaters to the border, the state's waters have plenty of muscle. That muscle would now be put to use in altering the course of hydroelectric power and its development.

Debts have been the cause of many desperate deeds, but they seem an unlikely catalyst for such a momentous occasion as the world's first successful long-distance transmission of a commercial supply of electrical energy. In the late 1880s, the debts of the Gold King Mine at Ames in southwestern Colorado were mounting. The biggest drain on the funds was the huge task of hauling coal to burn in steam-generators, coal brought in by mule train from mines high above timberline and several

When Captain John Gunnison came to the Uncompahgre Valley in 1853, only cactus and sagebrush grew there. But after the 1909 completion of a 5.8-mile tunnel bringing water from the Gunnison River, farms like these in Ouray County near Ridgeway became some of the richest in the Rocky Mountains. JAMES FRANK.

valleys distant. In a last-ditch effort to eliminate that expense, the owners of the mine turned to the Lake Fork of the San Miguel River.

In 1891, a generator was assembled by Lucius Nunn and George Westinghouse in a slate-board structure beneath a cirque wall where the Lake Fork tumbled over a high headwall. From Trout Lake above the headwall, the river dropped 320 feet, gathering speed as it fell and spinning the six-foot turbine of the experimental generator. Like a wild horse being ridden for the first time, the river nearly bucked the system off its hinges. The 3,000 volts originally anticipated became 10,000 volts and the crew had its hands full of sparks, shooting them through wires as alternating current three miles up and over the mountains to the site of the Gold King Mine. Despite a fear of lightning which could short out the whole system in an instant—and scenes like the one in which an operator standing atop a huge wax-soaked wooden block swept away a jagged arc of errant current, strung like a tiny lightning bolt between jacks, with his felt Stetson to prevent short-circuiting—the project was a success and stands today as the forerunner of modern hydroelectric power in the state. The rivers had been turned to sparks.

With the barrier of long-distance transmission crossed, the sights of hydroelectric engineers turned to an even more formidable barrier, the Continental Divide. In 1909, deep in Glenwood Canyon which, although the railroad had penetrated its depths, was still rugged and wild country, another engineering challenge was being met. One of the first dams on the mainstem of the Colorado River, at that time still known as the Grand River, was the Shoshone Dam just upstream from Glenwood Springs. Besides the gigantic technologic problems encountered by the project, proponents met head-on with one of the first public oppositions to the destruction of a river's beauty.

The plan to plug up the Colorado River for electricity enough "for the entire state" was opposed by a growing group of conservationists from Denver with the backing of the White River National Forest. The canyon's beauty was worth more to the human spirit, they said, than all the light bulbs in Denver.

In one of the earliest mitigation agreements, the project was altered to consist of a 2.3-mile tunnel bored into the canyon walls which would take the flow of the river diverted by a smaller dam and direct it through the turbines underground before the river was returned to its course downstream. To transport the man-made lightning, the first transmountain 100,000-volt transmission line arced across the Divide and sparks went out over mountains.

Despite such a grand start, hydroelectric power has never reached its potential in Colorado, having been overshadowed by the power pent up in huge native deposits of natural gas and coal. Only twelve to fifteen percent of the state's total electrical output is generated by its thirty-one hydroelectric stations, the largest of which are the triple units of the Curecanti Project on the Gunnison River (Blue Mesa, Morrow Point, and Crystal power plants) which have a combined generating capacity of 208 megawatts. Another three hundred sites have been surveyed for hydroelectric potential by the U.S. Department of Energy, and sixty-six of the sites proved suitable for major projects while a large number of the rest were slated as possible locations for the smaller "Low-Head Hydro" units which have recently been developed. Yet, many of these sites are located on various stretches of the eleven state rivers under study for inclusion in the National Wild and Scenic River Systems and, without a major energy crunch in Colorado, the power of water is seen by many people of the state as more beautiful when played out in the walls of a canyon than in a tangle of electric wires.

Correcting nature's errors

There is a fundamental imbalance in Colorado. The waves of settlers who populated the state swept in from the East and were stopped short by the flanks of the mountains. They settled the dusty plains of eastern Colorado as their home. The waves of stormclouds which sweep into Colorado from the west are stopped short by the western flank of the mountains. The precipitation settles on sparsely populated valleys and flows back towards its home in the Pacific, unseen by the dry cities of the Front Range. The West Slope, with 37 percent of Colorado's land area and just 20 percent of its population, flows with 69 percent of the surface water in the state. The Front Range, with 80 percent of the people and 63 percent of the land, has only 31 percent of the water.

After the rivers of the South Platte Basin had been diverted to the bone after all the reservoirs which could hoard water for the summer were built along its course, the people of the Front Range looked upon this imbalance as a mistake of nature and set out to right the wrong. Because of their higher population, which translates into greater political power, and because the Front Range settlements had sent engineers out over the Divide early to lay claim to what water they could, the time had come to put that water to beneficial use. By now, water was flowing under the mountains of the West Slope in projects like the Gunnison Tunnel. Electricity was flowing over the mountains of the Continental Divide. The next logical step was to bring water under the Divide itself to where, in the eyes of those on the sunrise side of the mountains, that water belonged. This next step was the Colorado-Big Thompson project.

Like a huge boulder rolling down the eastern side of the Divide, the proposed project began slowly and without opposing forces and obstacles to overcome. The proposal was epic in proportion—collect the waters of the Colorado River and its tributaries high upstream near their headwaters; store them in a triple lake system; drill the longest tunnel in the world beneath the Continental Divide; feed the water through a series of power plants on the East Slope; collect it again in another series of reservoirs; and dole it out to irrigators along the South Platte River. It was a project to rival the pyramids. In a proposal so complex that few admitted to fully understanding its import, Congress authorized the project in 1937.

But there were still other hurdles besides Congressional approval to clear before the task could be started. The citizens of the West Slope were not thrilled about the prospects of losing the proposed 390,000 acre-feet of water which would be siphoned annually out of the Colorado Basin by the giant straws of the project. After delicate negotiations, a mitigation plan was agreed to in which the Front Range would foot the bill for a water storage facility on the Blue River in exchange for the cooperation of the West Slope in completing the Colorado-Big Thompson Project. Green Mountain Dam and Reservoir northwest of Dillon were born.

Next, the project was opposed by conservation groups such as the Wilderness Society and the Izaak Walton League. At stake was the integrity and puri-

Colorado-Big Thompson Water Project

The Colorado-Big Thompson Project, consisting of more than 100 structures integrated into a transmountain water diversion system, supplies water for more than 700,000 acres and 400,000 people in the South Platte River Basin. It also generates 760 million kilowatt-hours of hydroelectric power, furnishing the electricity needs of the project system itself as well as those of customers in northern Colorado, eastern Wyoming, and western Nebraska.

Water of the Colorado River and Willow Creek is gathered in Lake Granby, pumped uphill into a canal, and channeled first into Shadow Mountain Lake and then into Grand Lake. From there it flows directly into the Alva B. Adams Tunnel and 3,780 feet below the Continental Divide into Marys Lake, then farther by tunnel into Lake Estes. This West Slope water can be diverted south to Carter Lake and tributaries of the South Platte River, or north to the Big Thompson River, Charles Hansen canals, and Poudre River, where it also flows into the South Platte River.

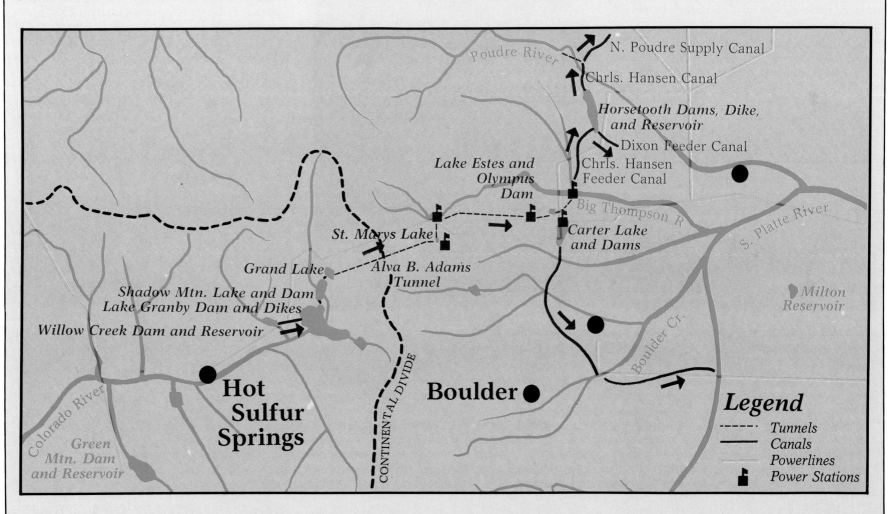

ty of Rocky Mountain National Park, given protection just twenty-two years earlier in 1915. The path of the tunnel would pierce the heart of the Park. A plastic model of the completed project was showcased around the state by Bureau of Reclamation to prove that the outlets of the tunnels would be located outside the boundaries of the Park and that no surface disturbance would occur. State and national pride in the Bureau of Reclamation still ran high because of its past successes, such as the Gunnison Tunnel; and the men who were, in the public's eye, what author David Lavender has called "white knights in rubber boots," were given the benefit of the doubt. The project was given the green light, and now only the Rocky Mountains stood in their way.

In 1940, work began on the Alva B. Adams Tunnel, named in honor of the Colorado senator. Over the next four years, two drilling teams inched towards each other from opposite sides of the Park. On June 10, 1944, the two sides met, and the tunnel—13.1 miles long; nine feet, nine inches in diameter; 3,780 feet below the surface of the earth—was opened. Three years later, on June 23, 1947, the first waters flowed through the tunnel and into the Big Thompson River.

The tunnel is the star of the project, but there are over one hundred other engineering components which make the whole system click, including ten reservoirs, thirteen dams, seven power plants, four pumping stations, twenty-four tunnels, and eleven canals. By 1954, most of the system was on line. Water from the Colorado River and its high tributaries is gathered in Lake Granby on the West Slope. From there, it is hefted 125 feet to a canal at Shadow Mountain Lake and into Grand Lake, where the surface level has been lifted by means of a dam so that gravity will feed the water into the tunnel. As it reaches the East Portal of the tunnel, it is still 2,900 feet above its final resting place in the fields of the South Platte Valley, and the energy of the water falling over that drop is used in a series of power plants to produce 760 million kilowatt hours

Grizzly Reservoir on Lincoln Creek is one of more than 1,900 reservoirs which dot the Colorado landscape.
KAHNWEILER/JOHNSON.

of electricity which is then marketed to eleven cities and nine rural utility cooperatives to serve 324,000 people. In addition, all the power needed to run the entire system is produced with a 69,000 volt powerline encased in a tube of nitrogen gas which runs inside the tunnel, connecting the project units on both sides of the Divide.

Yet water, not electricity, was the point of the Colorado-Big Thompson Project. After all the electrical power has been drained out of it, the water is again cupped into a series of storage units such as Carter Lake and Horsetooth Reservoir. From there, it is diverted to the irrigation ditches of 700,000 acres of sugar beets, alfalfa, corn, and potatoes in the South Platte Valley and used to supplement the water supplies of thirteen communities. The planned yield of 390,000 acre-feet annually has been cut down by such factors as compact stipulations to an average annual flow through the tunnel of around 250,000 acre-feet.

The first of that flow showed up in the irrigation ditches of the South Platte Valley in the nick of time. As a drought choked the South Platte Valley in 1954, a flow of 300,352 acre-feet through the tunnel came to the rescue. Estimates say that without that shot in the arm from the Colorado-Big Thompson Project, $22 million of the harvest of 1954 (which was valued at $41 million) would have been lost. Twenty-three years later, when another drought hit the Front Range in 1977, a flow of 309,477 acre-feet from the Colorado-Big Thompson provided an estimated fifty percent of the irrigation water used that year to produce a crop valued at $270 million. Proponents of the project point to these two banner years and the combined total of $154 million in crops saved as proof that the $160 million Colorado-Big Thompson project did indeed correct one of nature's grandest errors.

The success of the Colorado-Big Thompson Project spawned other tunnels and ditches which pierced the Continental Divide. Currently, eight tunnels and sixteen ditches coax water bound west to flow towards the sunrise side of the mountains, bringing an annual flow of some 630,000 acre-feet of "misplaced" Pacific-bound water to the Rio Grande, Arkansas, and South Platte rivers. But the source of the flow may be drying up as far as any additional transmountain diversion projects are concerned.

Proponents of recent plans for drilling even more holes in the already honeycombed Divide have met with obstacles which may prove even more formidable than the mountains themselves—the growing political power of the West Slope has led to a tighter grip on the flow of rivers on that side of the Divide, with projects constructed or planned which would keep the rivers at home. Besides, prime locations for additional transmountain diversions are difficult to find. The best sites have already been used, leaving only marginal routes which would make projects under existing technology too costly or difficult.

The economic picture for such large water projects has also changed, in two ways. First, inflation and construction costs make even projects with potential for a high return a strain on the local economy. The star of the system, the Colorado-Big Thompson Project, despite its paybacks, ran 347 percent over its original cost estimates. Current projects like the McPhee Dam on the Dolores, estimated in 1966 dollars to cost $46 million and with a price-tag in 1985 dollars of $452 million and climbing, have raised serious questions about the fiscal responsiblity of the Bureau of Reclamation and its projects.

In the face of such questions, the federal pocketbook slammed shut in 1977 with the "hit list" of President Carter. On the list were eight western water projects, including five in Colorado such as the McPhee Dam on the Dolores and the Narrows Dam on the South Platte, which President Carter pointed to as either uneconomical or environmentally destructive. More important, the list served notice that Uncle Sam would no longer be the money-tree for such projects. Although Carter's loss of western voters to Ronald Reagan in the 1980 election was, in part, due to his opponent's pledge of support for western water projects, continued budget restraints, the transfer of federal programs to the state, and ineffective agency management have kept the flow of federal money at a trickle.

So that some may flow free

There is a spot, deep in the northwest corner of the state, where two rivers come together. It is a gravel bar of willows and grey, river-worn rock, with heron tracks in the mud. At that spot, it is possible to trail your hands in two rivers at once, the Green from the north and the Yampa from the east. It is slickrock country and both rivers carry the panned-out colors of their canyons. At certain times of the year, both rivers flow with the same depth, exactly doubling where they meet. Both rivers carry deep and shadowy canyons on their backs. But there is a difference, and with your hands in the waters it can be felt like a pulse. The Yampa is a wild river; the Green is not.

In 1950, the U.S. Bureau of Reclamation proposed a 525-foot-high cement dam for the confluence of the Green and Yampa rivers in Echo Park, the heart of Dinosaur National Monument. The dam would have backed up the two rivers for forty-four miles. STEWART M. GREEN.

In the summer of 1954, few people had heard the term "wild river," and fewer still had heard of Echo Park at the confluence of the Green and Yampa rivers. Both were about to become the talk of every campfire, at the center of a controversy which would become a watershed in the American conservation movement.

In the early 1950s, the Bureau of Reclamation began prodding Congress to authorize their ten-dam, billion-dollar package called the Colorado River Storage Project. The long and bitter battles which had led to the division of Colorado River waters between the Lower Basin states (California, Arizona, and Nevada) and the Upper Basin states (Colorado, Utah, Wyoming, and New Mexico) were thirty years behind now. The new agreement of 1948 which further clarified the split of water between the Upper Basin states was signed and delivered. It was time that this water was put to good use, and the Colorado River Storage Project was the Bureau of Reclamation's idea of good use.

The plan included four dams to be built in Colorado: Morrow Point, Blue Mesa, and Crystal, all on the Gunnison, and one—Echo Park—at the confluence of the Green and Yampa rivers, within the boundaries of Dinosaur National Monument. Still stinging from the loss of the Tuluoumne River in California's Yosemite National Park to the Hetch Hetchy Dam, the proposal to dam the Yampa and Green within a national monument struck a raw nerve with conservationists. Few, if any, of those who stood up against the damming of Echo Park were even born when the Hetch Hetchy Project was first proposed in 1882, and most were only children when it received approval in 1913. Still, the fight for the heart of Yosemite National Park, led by John Muir (who founded the Sierra Club, which played such a large part in the Echo Park controversy), was well-known among the growing numbers of the conservation movement. The fires that had burned in Muir and others over Hetch Hetchy were rekindled by the fight for Echo Park.

In many ways, the Echo Park controversy proved even more important than the earlier battle. It was a measure of the strides made since Hetch Hetchy in understanding the concept of wilderness and wild rivers.

The controversy required that a nation articulate what it felt deep in its gut: that some places and rivers must remain wild. And it was a test case for other projects being proposed within the boundaries of places like Grand Canyon National Park, Adirondacks State Park, and Glacier National Park. If Echo Park had fallen, the others would have followed right behind. The stakes were high and both sides understood them.

The battle lines were drawn, but in this case they were not always so clear. Echo Park was not a classic, clear-cut confrontation. The 1950s were a time of growth and development, but that was not the issue. Even the staunchest of conservationists made it clear that this was not a matter of opposing the growth of a nation, but a matter of supporting the wildness of its parks and monuments. The question was not *whether* the dams were to be built, but *where*. That point sometimes made for confusing alliances and, in the end, nearly cost the country a wild canyon.

As the first trickles of the 1955 spring run-off began to stir in the Yampa River, debate on a bill authorizing the Colorado River Storage Project was drawing to a close in the Senate. A last-minute amendment was offered which would have deleted the Echo Park Dam from the bill, and support for such an amendment had emerged at public hearings; but that support was not unified enough to be effective and the amendment was voted down. When the final Senate vote was tallied, the Colorado River Storage Project, complete with the Echo Park Dam, had been passed.

The bill still faced debate and a vote in the House of Representatives before it could be signed into law by the President. And like the Yampa River, slowly rising with the winter snows melted by a spring sun, public opinion against the project was mounting under continued heat from conservationists. It was during these critical months, while the House was preparing to vote on the issue, that the country found its voice.

The most visible spokesmen for the conservationists were people like David Brower, then the executive director of the Sierra Club, and Howard Zahniser of the Wilderness Society. But their voices were given strength by support from all corners of the nation. It was as if a country of people, struggling to articulate a belief they had never before been so urgently asked to defend, suddenly found a common voice in defense of a remote slickrock canyon called Echo Park.

A flash flood of letters was unleased on Congress, running eighty to one in favor of keeping the dams out of Dinosaur National Monument. Books, like *This is Dinosaur*, edited by Wallace Stegner, came to the defense of the Monument; and where language failed, the photography of people like Phillip Hyde and Elliot Porter gave shape to feelings too deep for words.

By 1956, the tide had turned. Although nearly every member of the U.S. Congress was still in favor of the Colorado River Storage Project, the Echo Park clause had been transformed into a glaring political disadvantage. So when H.R. 3383, a bill to authorize the Colorado River Storage Project, was voted on and passed by the House of Representatives, an amendment had been added stipulating that ". . . no dam or reservoir constructed under the authorization of this Act shall be within any National Park or Monument."

The Echo Park Dam was scrapped; another Hetch Hetchy was avoided. If the vote had gone differently, the gravel bar at the confluence of the Green and Yampa rivers, with its heron tracks, would have been drowned beneath a forty-four mile reservoir. Flaming Gorge Dam was built upstream on the Green River beyond the north boundary of the Monument, and the three dams of the Curecanti Project on the Gunnison have drowned all but thirteen miles of the fifty-three-mile Black Canyon of the Gunnison; but the Yampa River still flows free past the gravel bar in Echo Park, the place which gave a conservation movement its voice.

Headwaters of a system

About the time the vote on the Echo Park controversy was being tallied in Washington, two men were rowing their rafts down Montana's Middle Fork of the Flathead River. Over a morning campfire, John Craighead and Clifton Merritt discussed their ideas for establishing a "natural rivers system," a program, like the National Parks, which would offer protection for some of the nation's remaining free-flowing rivers. After the trip, Craighead, who was to become a well-known wildlife biologist, would write up his ideas for the *Naturalist* magazine. Merritt, now the executive director of the American Wilderness Alliance, would begin the task of gathering support for a wild river system among citizens and political leaders.

But the idea seemed to get caught in an eddy. Although the Echo Park victory had given conservationists new life and introduced the world to the term *wild river*, all focus was now on a different system, the wilderness system which would culminate in the 1964 passage of the Wilderness Act. It was four more years before eyes began turning to wild rivers.

On October 2, 1968, Congress passed the Wild and Scenic Rivers Act establishing the National Wild and Scenic Rivers System. Enactment of the new law was, as stated in the bill's preamble, a declaration that "the established national policy of dams and other construction at appropriate sections of the rivers of the United States needs to be complemented by a policy that would preserve other selected rivers or sections thereof in their free-flowing condition. . . ."

It was a declaration of limits, and that a balance needed to struck. Some rivers, it said, should flow free.

With a stroke of the pen, eight rivers—from the Rogue in Oregon to the Wolf in Wisconsin—became the foundation of the National Wild and Scenic Rivers System. In addition, twenty-seven other rivers like the Allegheny in Pennsylvania and the Middle Fork of the Flathead in Montana, where Craighead and Merritt had first spoken of the need for such a system, were authorized for study and possible designation. None of the rivers in Colorado were named in the original bill, but the door had been opened and hopes for protection of wild rivers like the Los Pinos, Piedra, and Yampa ran higher than spring floods.

There were dreams of a wild and scenic river system cradling a hundred rivers in its protective arms by 1978, and double that number by 1990. That dream is not coming true, however. Progress in the system has been slow. The number of rivers protected moved forward so slowly at first that the system seemed to be rowing against a strong headwind. The four years directly following the 1968 passage were drought years for the system with not a single river added. Since that time, things have seemed to meander between flurries of activity and logjams. One such flurry came on January 3, 1975 when an amendment was passed that added twenty-nine rivers to the list of those to be studied and of that twenty-nine, twelve flowed in Colorado. The stage seemed set.

The numbers game

If numbers alone could tip the cart, Colorado would have its wild rivers. A recent study of whitewater boating in the state showed that river use from 1976 to 1982 increased at a 9.5 percent annual rate to 255,377 user days in 1982 (a user-day equals one twelve-hour period). Given that annual rate of increase, river use for whitewater boating alone in 1985 is nearing 300,000 user days, with the most popular boating stretches being the Arkansas and the Upper Colorado. The study, by the Western River Guides Association, also estimated that more than 200,000 people paddle a kayak or row a raft on Colorado rivers every year.

Those numbers translate into money for the state's economy. Fees are paid guides by those taking commercial trips; boats, oars, paddles, and lifejackets are bought or rented for the trip. Such direct expenses total $14 million a year in Colorado.

But that is not all, the study showed. Gas must be pumped into cars used for getting there and

Colorado's Gold Medal and Wild Trout Rivers

Wild Trout Rivers

River	Designated Section	Mileage
North Platte	Routt National Forest to Colorado/Wyoming line	5.3
Laramie	Within Hohnholz State Wildlife Area	2.5
Poudre	Monroe Tunnel to Poudre Valley Canal headgate	4.3
Poudre	Hombre Ranch to Granpa's Bridge	4.2
Poudre	Poudre Fish Hatchery to confluence of Black Hollow	4.7
South Platte	Beaver Creek to South Platte Arm Gauging Station	9.0
South Platte	Cheesman Dam to Wigwam Club	3.0
Arkansas	Gas Creek to Four Mile Creek	6.0
Arkansas	Texas Creek to Parkdale	14.0
Conejos	Menkhaven Ranch to Aspen Glade Campground	4.0
Colorado	Gore Canyon to State Bridge	16.0
Fraser	One mile below Tabernash to one mile above Granby	5.0
Blue	Green Mountain Dam to 2.5 miles downstream	2.5
Roaring Fork	Hallum Lake to Woody Creek bridge	7.0
Gunnison	Black Canyon to North Fork	26.0
East	Roaring Judy Fish hatchery to Taylor River confluence	5.0
		118.5

Gold Medal Trout Rivers

River	Designated Section	Mileage
Arkansas	Stockyard Bridge in Salida to Fern Leaf Gulch	28.0
Blue	Green Mountain Dam to Colorado River	16.0
Colorado	Windy Gap Reservoir to Troublesome Creek	20.0
Fryingpan	Ruedi Reservoir to Roaring Fork River	14.0
Gunnison	Black Canyon to North Fork of the Gunnison	26.0
Rio Grande	Coller Wildlife Area to Farmers Union Canal	22.5
Roaring Fork	Crystal River to Colorado River	12.0
South Platte	Cheesman Dam to North Fork of South Platte	19.5
		158

back; motel rooms are welcome the night after the trip; a souvenir should be brought home for grandmother. Such indirect expenses add another $24 million and bring the total yearly boost to the state's economy from whitewater boating to $38 million.

And water, as any good flyfisherman knows, is meant as much to fish as to float. There are eight thousand miles of trout waters in Colorado, including 158 miles of gold medal waters designated as trophy-class fisheries, and 118.5 miles of wild trout waters where anglers are tested against naturally reared rather than hatchery-bred stock. In 1984, the Colorado Division of Wildlife sold 759,925 licenses to fishermen, bringing the state a direct revenue of $8,263,260.

These are the simple calculations. It is fairly basic arithmetic to multiply the number of fishing licenses sold in the state by the cost of those licenses, but that is not the full value of a river. It is simple to add up the money spent by 100,000 people who floated the Colorado River within the state or by 75,000 who boated the Arkansas, but that sum provides only the economic value of whitewater boating, not the true economic value of a free-flowing river.

For decades, however, these figures have been the only ones available. Proponents of wild and scenic rivers were forced to battle the cold, hard numbers of dam-builders and benefit-cost ratios with just a few scant facts about the economic value of recreation and a conviction that there was inherent worth in a river *as a river*, flowing free—not measurable in dollar signs. There has been a lot of rhetoric for decades about "priceless resources" and "invaluable" assets, but until recently no one was able to insert a decimal point.

Rivers are more than recreation days and fishing licenses, more than a stream of benefit-cost ratios and projected kilowatt hours—much more. A 1985 study by Colorado State University's Department of Agriculture and Natural Resource Economics used the eleven rivers in Colorado recommended for inclusion in the National Wild and Scenic Rivers System to prove it.

When Lawn Lake Dam in Rocky Mountain National Park gave way in 1982, flooding downtown Estes Park, three people died and $30 million in property was lost. A subsequent study by the state water engineer showed many of the state's 2,200 dams in urgent need of repair if similar incidents were to be avoided. JAMES FRANK.

The study, building on a framework of concepts pioneered by noted natural resource economist Dr. John Krutilla of Resources for the Future, attempted for the first time to attach a dollar figure to the intangibles of a wild river. The study team called these factors "preservation values" and listed them as three types: the value of knowing rivers will be wild for potential visitation (option value); the value in knowing that wild rivers exist as habitat for fish and wildlife (existence value); and the worth of knowing that rivers will be flowing wild and free for future generations (bequest value). Using the "willingness-to-pay" method proven valid by such programs as the Non-game Check-off used on Colorado State Income Tax forms to provide money for the Colorado Division of Wildlife, the research began to discover where that elusive decimal point should go.

The average Colorado household responding to the study indicated a willingness to pay up to $95 annually for both the recreational and preservation benefits of the eleven rivers. By multiplying that figure by nearly 1.2 million households in Colorado, the researchers estimated that the total annual benefit of preserving the eleven rivers was $112.6 million. Using a standard technique of economics called discounting, the same practice followed by proponents of water projects when figuring out a benefit-cost ratio, the present value of designating the eleven Colorado rivers as components of the National Wild and Scenic Rivers System was $1.4 billion.

However, there is certainly another side to this ratio—costs. In the case of rivers, the study team found, there are three main cost factors involved in designation: management of the river; operating costs for the system; and the value of any foregone water projects prohibited under the Wild and Scenic River Act. Of the many dam projects proposed for the eleven rivers, all but two are, for unrelated reasons, unlikely to be built and so pose little or no cost as projects foregone because of designation. The value of those dam projects which *would* be cancelled by river preservation, added to management and operating costs, and discounted at the same rate as benefits were, produced a total of $15.8 million in costs.

The final computation, to arrive at a benefit-cost ratio of the kind so often argued by proponents for water projects, puts the figure at 95:1—in other words, benefits outweigh costs nearly a hundredfold. By comparison, the benefit-cost ratio of the Colorado-Big Thompson Project when proposed was just 12:1. At that rate, designation of the eleven rivers seems like a bargain and the study showed that four out of five respondents to the survey said they would pay to see that designation granted.

Even these numbers, however, do not tell the whole story of a river's worth. There are other treasures of a wild river which will never be chained to a decimal point or cooped up in a benefit-cost ratio. There is no measure for artistic inspiration, the beauty of a slickrock canyon gone gold with sunset. There is no measure to the value of silence as deep and soft as moss, no price for the whisper of wingbeats on the water as night birds rise. The university study is less an attempt to hang a price tag on each of our remaining wild rivers than it is a step towards validating a feeling that all of us have: that a free-flowing river carries a value beyond the current price of an acre-foot of water or the daily rate for a commercial raft trip. It is an attempt to quantify some important factors but not all. As for those other values, of sunsets and wingbeats, *priceless* is still the only word appropriate.

The Colorado logjam

There are places, deep in the Gorge, where the Gunnison River still runs wild. From a rock ledge overlooking the river where it is pinched to rapids, the air of a side canyon is still suddenly ripped with the sound of a blade drawn across stone, and a peregrine falcon cuts into the light of the canyon, searching the wind for its angles, and then slants off downstream. The sight of a peregrine, like the howling of wolves, is a sign of a pure and untamed place. At least one section of the Gunnison is still wild.

But for the Gunnison and the ten other Colorado rivers which were recommended for addition to the National Wild and Scenic Rivers System, "wild" is still just a word heard in the glow of a campfire. Of the twelve rivers initially authorized for study in the 1975 bill, only the Big Thompson did not show the "outstandingly remarkable" attributes necessary to wear the official title of "wild river." Yet Colorado, which has more miles of wild river within its borders than any state outside of Alaska, has not added a single mile to the National Wild and Scenic Rivers System.

Nationally, the federal system currently protects 7,217 miles of sixty-five rivers. None are in Colorado. Twenty-eight individual states, through state wild river programs initiated since Wisconsin began the first in 1965, protect another 11,404 miles of 318 river segments. Again, none are in Colorado. The reason for this wild river drought in Colorado can be best described as a political logjam.

The biggest tangle in the logjam is the Colorado River. Life in the West, as so long ago recognized by explorers like Long and Powell, is held together by the braids of rivers which bind the plains and the mountains like long ropes. To let the rivers flow downstream without tapping them has been seen, in some eyes, like letting go of the rope. Nowhere is this attitude and its effect more clear than in the long history of battles for the thickest rope in the West, the Colorado River.

As early as 1919, all seven of the states through which the Colorado River flows looked out across the dry land and came to the same conclusion: water is the key to life in the West, and the Colorado River is the key to water. Everyone wanted to be dealt in. So in 1922, representatives from those seven states met in Sante Fe, New Mexico

Introduced variety

Despite the wide variety of aquatic habitat presented by the state's rivers, falling from high mountains to plains, few native fish species have evolved. In the Colorado River Basin, less than ten native species evolved and seven of these are found in no other waters on earth. Fishing in Colorado is an introduced sport.

Still, fishing creates the highest number of recreation days of any wildlife-related sport in Colorado, including big game hunting. Over three-quarters of a million fishing licenses are sold each year, and these anglers account for 23 million fish caught annually.

Cold water species, such as trout, fill the biggest part of Colorado fishermen's creels, and by far the most popular trout is the rainbow. Introduced to Colorado waters in the 1880s, this colorful trout native to West Coast streams has found an ideal home in the riffles and backwaters of Colorado rivers. Each year, millions of eight-inch rainbow trout are stocked in rivers, streams, and reservoirs across the state and account for 75 percent of all the fishing success in Colorado. The two other popular trout species, the brown trout and brook trout,

Fishing in Colorado is an introduced sport. Each year, millions of fingerling rainbow and brook trout are raised in hatcheries across the state and released into Colorado's fishing waters. ROBERT PITZER/AMWEST.

are also introduced species. Rivers like the Roaring Fork, Poudre, upper Arkansas, Gunnison, and Los Pinos are among the most popular fishing waters in the state, attracting anglers from around the world to test the waters of Colorado rivers.

Wild rivers aren't just for people. Here a cougar, in the warm light of dusk, drinks from one of the thousands of streams which provide vital habitat to numerous species of wildlife. BRIAN PARKER/TOM STACK AND ASSOCIATES.

However, Colorado recreationists love their wild rivers. Each year, for example, whitewater competitions known as "river rodeos" are held on rivers like the Animas near Durango. SCOTT S. WARREN.

and began to divide the pie. Sante Fe seems a long way from the banks of the river for such a meeting, and perhaps that played a part in the problems which plagued the resulting agreement.

After looking back on a few years of water records (too few, it would turn out), the panel set the annual flow of the Colorado River at 20.5 million acre-feet per year at gauging stations near the mouth of the river. An imaginary line was drawn near the mouth of the Paria River just upstream from the Grand Canyon. Everthing downstream of that line became known as the Lower Basin and included California, Arizona, and Nevada. Everything upstream of the line—Colorado, New Mexico, Utah, and Wyoming—became known as the Upper Basin. Each basin was then allotted 7.5 million acre-feet of the Colorado River every year, leaving a "safety margin" in the event that Mexico should, as it did in 1944, demand its 1.5 million acre-feet share of the river. At the time of the 1922 Colorado River Compact, no attempt was made to divide the water further among the states of each basin; it was enough that the river itself had been parceled out, to the tune of 16.5 million acre-feet every year.

An acre-foot of water is equivalent to 325,851 gallons, enough to flood one acre of land one foot deep, and thus the name. A single acre-foot is enough to satisfy the drinking needs of four average-size families for an entire year, flush 65,170 toilets, or fill 25 swimming pools and several large hot tubs. With 16.5 of those acre-feet expected out of the Colorado River annually, the trouble began.

Some simple arithmetic illustrates the problem. When the river was divided up in 1922, the West was dripping in an uncharacteristic wet cycle. More in keeping with its arid reputation, the Colorado River averaged only 13.1 million acre-feet from 1929 to 1979. Subtract the 16.5 million acre-feet of water asked of the river, from the 13.1 million acre-feet the river usually has to give and the problem is clear—the Colorado River is overbooked. Just below Imperial Dam, the last of the sixty-five dams on the river in the Lower Basin, the rope snaps. Out on a bone-white, salt-thick flat sixteen miles from the river's natural mouth at the Gulf of California, the Colorado River sinks into the sand and stops.

It is a sight that no one living in the West should escape: the sight of a river as powerful as the Colorado, the river that cut the Grand Canyon and drains one-twelfth of the entire United States, flowing thick as spit and then dropping into the

This burglary is about to come to a quick and dramatic end. In 1986, the $3 billion straws of the Central Arizona Project will begin to suck up the water for Phoenix and Tucson, leaving the taps in California dripping dust. Already, a full-scale Water War has begun.

All of this may seem like a downstream problem, but it's not. One solution to keeping the taps of San Diego flowing, after the Central Arizona Project is put on line, is a program of leasing. States like Colorado, which have allotted rights to water not currently being put to use, could lease their excess to California until a more long-term solution to the problem can be found. Eight of the eleven rivers in the state which have been recommended for protection are within the Colorado River drainage—a leasing program could mean life for these rivers since they would serve as the natural transport system for any leased water. But questions remain concerning the legal implications of such water leasing programs, questions which must be resolved to the satisfaction of nervous Colorado water watchers before leasing desert like a spent pack horse. Meanwhile, the lawns of Denver are ankle-deep in Kentucky bluegrass and the golf courses of Phoenix are as green as hundred dollar bills.

The death of this river has been widely publicized, but instead of inspiring programs of conservation or more efficient water use, publicity has sparked a water grab reminiscent of the gold rushes, complete with claim-jumping. California, the state which contributes the least water to the system, gulps the biggest share. As states such as Colorado and Arizona grew more slowly, letting some of their allotted water from the Colorado River Compact flow downstream, California has helped itself, using the "surplus" water to fuel the growth of cities like San Diego. The water heating in the hot tubs of San Diego and washing cars in Los Angeles is, at least in part, water with names of Arizona and Colorado in it.

Storm clouds gather over the Elk Mountains in Gunnison National Forest, while this rocky stream near Schofield Pass slips smoothly on its way. WILLARD CLAY.

A modern raindance

It didn't include chanting or wooden divining rods, just a magic powder called silver iodide; but the goal of the Winter Orographic Snowpack Augmentation Program (WOSAP) was the same that man has dreamed about for centuries—controlling the weather. In a pilot program conducted by the Bureau of Reclamation in the San Juan Mountains between 1970 and 1974, an attempt to add water to Colorado rivers by artificially increasing snowpack flow brought both more snow and more concerns.

Results of the study indicate that a full-scale program could increase the annual snowpack in areas above 9,000 feet in elevation by 20-25 percent, adding approximately 2.3 million acre-feet of flow to the state's rivers. The extra water, proponents of the project say, could solve the water problems of the oil shale industry in the northwest part of the state and be used to fuel growth across the West.

But the pilot program experienced difficulty in targeting areas for added snow. Several times, seeded storms drifted out of the test area and dumped massive amounts of snow on nearby towns like Telluride, where the collapse of the jail roof was blamed on one such errant storm. Moreover, the environmental impact of a large-scale weather modification program could be severe—increased snow means increased potential for avalanches, a greater risk of spring flooding, the loss of vital winter wildlife range, disruptions of vegetation patterns such as the lowering of timberline, and the unknown effects of pumping over eight thousand pounds of silver iodide into the atmosphere each winter.

The results of the four-year pilot program are still being debated. Man is within reach of his dream of controlling the skies, but that dream, without restraint, could turn into a nightmare.

becomes politically practical.

Evidence of that nervousness can be seen on the face of a sandstone cliff near the end of Ruby Canyon. There, written in chalk-white paint, are the names of Colorado and Utah, and between those names is a white line. The sight has become a sore spot for Colorado water engineers, not because it defaces the cliff but because of the "leak" in the state's water system it symbolizes. Each year, through all of Colorado's rivers, ten million acre-feet of water flows out of the state. Of that figure, just over eight million acre-feet is required by one of the nine interstate compacts involving Colorado rivers. The rest, amounting to 1.5 million acre-feet, is water for which the state has no storage or diversion facilities, although by law it belongs to Colorado.

In an effort to more closely monitor this "leak," Colorado water engineers have turned to the heavens for help, using space-age dipsticks forty-eight inches tall and equipped with two solar-panel fins. From eighty-two locations on major rivers and streams, these electronic measuring sticks send information on such vital river statistics as flow, depth, velocity, and salinity to a satellite floating 23,500 miles above the Colorado skies—every fifteen minutes, twenty-four hours a day, 365 days every year. From the satellite, messages are beamed down to a computer at the offices of the State Water Engineer in Denver and then distributed to seven district water offices across the state. With this eye in the sky, every drop flowing in the state is accounted for, even if it can't be put to use for irrigation or drinking water yet.

The program of leasing, the mechanical wizardry, and much of the political logjam which has spawned them all are indications of fear. Colorado's worst fear is that the water which it can't yet put to use will continue to squirt from California or Arizona garden hoses and when—and if— the time comes when Colorado claims that as its own by compact, only Water War II will be able to pry it loose.

Yet the case of *Arizona vs. California*, involving the Central Arizona Project, should reduce those worries a bit. In that decision, the U.S. Supreme Court upheld the Colorado River Compact and subsequent laws which distribute the water as binding and valid, so that Colorado's name is indelibly stamped on even that 1.5 million acre-feet leaking from the state each year. The decision has helped pry loose the logjam which has kept Colorado out of the river protection picture.

In 1982, the Reagan Administration proposed a bill which would have given protection to several rivers in the West, including four in Colorado— the Los Pinos, Elk, Encampment, and Conejos. However, the bill contained "release" language which would have given individual states the power to drop a river from a national study. The bill would have added rivers in Colorado and, at the same time, put rivers nationwide in jeopardy. So the bill was not given the support of river conservationists. But it was a start.

In 1983, Congressman Hank Brown of Colorado

The river lifeline

Colorado rivers are more than just rocks and water. They are a lifeline for many of the state's wildlife species. Of the twenty-five native Colorado species currently listed as "threatened" or "endangered" by the state or federal wildlife agencies, at least fifteen have their fates tied directly to the fate of the state's rivers.

River otter once roamed every major river basin in the state, but fell victim to the traps of the fur companies of the 1800s. Even with this pressure, the river otter might still be a common sight along the river if not for the decrease in water quality due to the mining and the destruction of habitat caused by development. The last recorded sighting of a wild river otter was in 1906 along the Yampa River. A program of reintroduction was begun along rivers like the Gunnison in 1976, but this sleek creature of the water is still listed as endangered in Colorado.

A major reason for the demise of these species has been the introduction of sport fish such as rainbow and brown trout and several species of bass. These fish have out-competed such native species in a habitat shrinking due to water development. A reintroduction program in Rocky Mountain National Park has begun to stabilize some populations of the native species, but more quality habitat must be protected to secure the future of the cutthroat trout.

There are others—the great sandhill crane shot for food and robbed of nesting habitat when rivers were depleted for irrigation and threatened further by the oil shale development on the Yampa and White rivers; the bald eagle, reduced to only two nesting pairs in the state (although as many as six hundred to a thousand of the majestic birds winter along state rivers); the johnny darter of the South Platte River.

And the list goes on. For these species and many others which are tied with less obvious strings to aquatic habitat, the rivers of Colorado are much more than rocks and water. They are a lifeline to survival.

took another step. After seeing his first wild river on a raft trip sponsored jointly by the American Wilderness Alliance and the Western River Guides Association, Brown called together a coalition of river conservationists, agriculture representatives, water resource experts, and local citizens to hammer out a bill for the protection of the Poudre River in his district. The ripples of that bill are still on the water and, despite the fact that the Poudre has not yet been protected, the efforts of Hank Brown did open a valuable line of communication between interest groups.

There have been other bills, affecting the Gunnison and Encampment, and each time a bit more of the logjam has been swept away. No "wild" rivers flow yet in Colorado, but every time the issue is raised, around a campfire or in a congressional hearing, the possibilities flow a little closer to reality. With each running of a river, more and more people come to see that a wild river is as rare and beautiful as the sight of a peregrine in the skies. And like the peregrine, which is on the federal endangered species list, wild rivers are endangered, too.

A wooden sign in Poudre Canyon tells the story of the dams planned for the Poudre River, dams which would inundate sections of the river now under consideration for protection as wild. JEFF RENNICKE.

Suggested reading

Wheat, Doug. *The Floater's Guide to Colorado.* Billings, Montana: Falcon Press Publishing, 1983.

Anderson and Hopkins. *Rivers of the Southwest.* Boulder, Colorado: Pruett Publishing, 1982.

Perry, Earl. *Rivers of Colorado.* American Canoe Association, 1978.

Redfern, Ron. *The Making of a Continent.* New York Times Books, 1983.

Farb, Peter. *Face of North America.* New York: Harper and Row, 1963.

Chronic & Chronic. *Praire, Peak and Plateau.* Denver, Colorado: Colorado Geologic Survey, 1972.

Chronic, Halka. *Roadside Geology to Colorado.* Mountain Press Publishing, 1980.

Griffiths and Rubright. *Colorado: A Geography.* Boulder, Colorado: Westview Press, 1983.

Stegner, Wallace, editor. *This is Dinosaur.* New York: Alfred A. Knopf Inc., 1955.

McTighe, James. *A Roadside History to Colorado.* Boulder, Colorado: Johnson Books, 1984.

Powell, J. W. *The Explorations of the Colorado River and its Canyons.* New York: Dover Press, 1961.

Stanton, Robert Brewster. *Colorado River Controversies.* Boulder City, Nevada: Westwater Books, 1982.

Waters, Frank. *The Colorado.* New York: Rinehart & Winston Publishing, 1946.

Dellenbaugh, Fredrick S. *The Romance of the Colorado River.* G.P. Putnam, 1902.

Hart & Hulbert. *Zebulon Pike's Arkansaw Journal.* Colorado College Press, 1932.

Williams, Albert N. *The Water and the Power.* Duell, Sloan & Pearce, 1951.

League of Women Voters. *Colorado Water.* Denver: League of Women Voters, 1972.

Walsh, Sanders & Loomis. *Wild and Scenic River Economics.* Denver: American Wilderness Alliance, 1985.

Lamm & McCarthy. *The Angry West.* Boston: Houghton-Mifflin, 1982.

Berkman & Viscusi. *Damming the West.* Grossman Publishing, 1973.

Colorado Water Conservation Board. *A Hundred Years of Irrigation in Colorado.* Colorado Water Conservation Board, 1952.

Tierney, Patrick, et. al. *The 1980 Colorado Whitewater Boating Use and Economic Impact Study.* American Wilderness Alliance and Colorado River Outfitters Association, 1980 and amended.

Tim Kelly's Fishing Guide. Hart Publication, 1983.

Collins & Nash. *The Big Drops.* Sierra Club Books, 1978.

An afterword:

Reflections on the never-ending story

The rivers of Colorado sustain life, but at a substantial cost. Walk the dry riverbed where the Rio Grande River, below Alamosa, has been drained dry by a series of pipes and canals. Smell the sour winds off the South Platte on a hot day where the water has been tainted with industrial wastes, chemicals, and the spoils of more than two million people who live along its course. Feel the confusion in tangled currents where water stolen from the West Slope by the Colorado-Big Thompson Project explodes from the darkness beneath the Continental Divide into the rivers of the East Slope. Or follow what remains of the Colorado River after more than sixty-five dams in its watershed, and feel the sadness of the great river where it crawls to a trickle miles short of its ancestral home in the sea. Imagine the way light played in grottos and how water echoed in deep holes along the forty miles of the Black Canyon now lost under the waters of the triple-dam Curecanti Project. The toll of life-giving sometimes feels like the dull ache of cold steel.

Much has been lost. Glenwood Canyon on the Colorado, which once rivaled any in the West, is now crammed with a railroad, four-lane highway, dam, and power plant. The Roaring Fork, clear as the sky, has a third of its water sucked into tunnels near Aspen and pulled under the Divide. The Blue River feeds through a twenty-four-mile straw to sprinklers on Denver lawns.

There is more. In a cruel twist of irony, a canyon on the Dolores River named Little Glen Canyon in memory of that place drowned behind Glen Canyon Dam in Arizona, is now being lost itself—not flooded, but emptied by the McPhee Project upstream. There is not a river in Colorado that is not, to some degree, diverted, channeled, tapped, or made to carry the burden of man's labors.

Of course, the rivers of Colorado do not belong solely to one state. Many of the state's rivers flow out beyond Colorado's boundaries to become for a time the rivers of Utah, or New Mexico, or some other state. Even those like the Gunnison or the Yampa or the Roaring Fork, which flow from headwater to mouth within Colorado, do not belong solely to the state. Each year, thousands of visitors from around the world step to the ledges of the Black Canyon of the Gunnison to witness the power of rock and water, or come to the Yampa when it is full with the thaw of spring to ride the whitewater. Hidden in the deep, sky-colored pockets of the Roaring Fork are the rainbow trout and reflected scenery which have made Aspen on the banks of the river an international resort even during the summer, when the winter snows flow off the slopes and into the river.

Yet for all their sacrifices, no river in Colorado is protected by the National Wild and Scenic Rivers Act. No river in the state has been given protection of any kind. Rivers are everywhere being dammed, diverted, and abused at a much faster rate than they can withstand. In Colorado, protection of wild rivers is now or never—there is hardly a stretch of river in the state not threatened by a proposal for a dam, irrigation project, pollution,

The rivers of Colorado are part of a dance between water and sky. Here on the Gunnison, the two mingle in the light of an early evening.
JEFF RENNICKE.

or other development.

Oil shale is the largest black cloud on the horizon for Colorado rivers. Although it lies dormant due to the economic climate, no one has forgotten the trillion gallons of recoverable oil lying in the black heart of the Piceance Basin in northwest Colorado. It seems inevitable that the economic outlook will change and with it the fate of rivers like the Yampa and the White which flow near the hidden oil reserves. At best, the delay in production will allow time for the development of new, less environmentally damaging techniques for recovering the oil.

But slowly, like the spreading of the glaciers which once crawled over the face of Colorado, the place of rivers in our lives is being recognized. Like the glaciers, such a realization could leave a major change on the face of the Colorado landscape.

The mood of the nation is different. No longer are expensive and environmentally devastating projects greeted with unquestioned flag-waving. Legislation such as the Wilderness Act, the Endangered Species Act, the Clean Water and Clean Air acts, and the Wild and Scenic Rivers Act provides evidence that we are beginning to recognize limits. These laws represent conscious decisions, by individuals and society, to protect areas of wild land and wild water, to preserve clean air and clean water, and to fight against the extinction of our fellow creatures.

Wild rivers—themselves an endangered species—belong as much to the future as to the present. The state of our rivers is a legacy which will be left for future generations. It is only by action now that the next generation will have the chance to know the sight and sound of a wild Colorado river.

Watching a river flow over the state line and out of Colorado does not bring the story of Colorado rivers to an end. In some ways, it is just a beginning. For the story of Colorado rivers has no end. Like the river at the border, it just flows on.

Reflections of the South Platte River as it slips downstream and continues the never-ending story of the rivers of Colorado.
MARK HEIFNER/
THE STOCK BROKER.

Quaking aspens at the foot of Mt. Sneffels in the San Juan Mountains. LARRY ULRICH.

*Coming up
in the Colorado Geographic Series:*

Colorado Mountain Ranges

Number two in the Colorado Geographic Series features the fascinating story of the mountains for which Colorado is so famous—both the famous, well-used ranges and the off-the-beaten-path, almost-unknown ranges. Like *The Rivers of Colorado, Colorado Mountain Ranges* will be richly illustrated with the same type outstanding color photography used in *The Rivers of Colorado*. The photos and graphics will blend with a fact-filled, easy-to-read text covering the history, discovery and development, recreational and economic use, and geology of the mountains of Colorado.

Colorado Mountain Ranges is scheduled for release in Spring 1986. Don't miss it.

Staying in touch... with the Colorado Geographic Series

You can turn this book into the start of a collection—*The Rivers of Colorado* is only the beginning. Falcon Press will continue the Colorado Geographic Series with the release of at least two books per year. All books in the series will be similar in format to *The Rivers of Colorado* and feature a new and equally interesting aspect of Colorado geography.

You can get advance copies of each book in the Colorado Geographic Series and receive a significant prepublication discount. Simply send us your name and address, and we will automatically send you an advance notice just before the release of each book. There is, of course, no obligation to buy any of the books. Send the notice to Colorado Geographic Series, P.O. Box 4368, Boulder, CO 80306-4368.

Extra copies of *The Rivers of Colorado* and future books in the Colorado Geographic Series are available in bookstores and other retail outlets throughout Colorado or by ordering directly from Falcon Press. To order, send $14.95 for softcover or $24.95 for hardcover, plus $1.50 per book for postage and handling, to Colorado Geographic Series, P.O. Box 4368, Boulder, CO 80306-4368.